Introduction to Mathcad® 11

Ronald W. Larsen
Montana State University

Pearson Education, Inc.
Upper Saddle River, NJ 07458

Library of Congress Cataloging-in-Publication Data
Larsen, Ronald W.
    Introduction to Mathcad 11
        p. cm.—(ESource—the Prentice Hall engineering source)
    Includes index.
    ISBN 0-13-008177-9
    1. Mathematics—Data processing. 2. Mathcad. I. Title. II. Series.

QA76.95 .L35 2000
510'.285'5369—dc21
                                                                        00-061131

Vice President and Editorial Director, ECS: *Marcia J. Horton*
Executive Editor: *Eric Svendsen*
Associate Editor: *Dee Bernhard*
Vice President and Director of Production and Manufacturing, ESM: *David W. Riccardi*
Executive Managing Editor: *Vince O'Brien*
Managing Editor: *David A. George*
Production Editor: *Craig Little*
Director of Creative Services: *Paul Belfanti*
Creative Director: *Carole Anson*
Art Director: *Jayne Conte*
Art Editor: *Greg Dulles*
Manufacturing Manager: *Trudy Pisciotti*
Manufacturing Buyer: *Lisa McDowell*
Marketing Manager: *Holly Stark*

 © 2004 Pearson Education, Inc.
Upper Saddle River, New Jersey 07458

All rights reserved. No part of this book may be reproduced, in any form or by any means, without permission in writing from the publisher.

The author and publisher of this book have used their best efforts in preparing this book. These efforts include the development, research, and testing of the theories and programs to determine their effectiveness. The author and publisher make no warranty of any kind, expressed or implied, with regard to these programs or the documentation contained in this book. The author and publisher shall not be liable in any event for incidental or consequential damages in connection with, or arising out of, the furnishing, performance, or use of these programs.

Mathcad is a registered trademark of Mathsoft Engineering & Education, Inc., 101 Main Street, Cambridge, MA 02142-1521.

Printed in the United States of America.
10 9 8 7 6 5 4 3

ISBN 0-13-008177-9

Pearson Education Ltd., *London*
Pearson Education Australia Pty. Ltd., *Sydney*
Pearson Education Singapore, Pte. Ltd.
Pearson Education North Asia Ltd., *Hong Kong*
Pearson Education Canada, Inc., *Toronto*
Pearson Educación de Mexico, S.A. de C.V.
Pearson Education—Japan, *Tokyo*
Pearson Education Malaysia, Pte. Ltd.
Pearson Education, *Upper Saddle River, New Jersey*

Engineering Success by Peter Schiavone of the University of Alberta intended to expose students quickly to what it takes to be an engineering student.

Creating Your Book

Using ESource is simple. You preview the content either on-line or through examination copies of the books you can request on-line, from your PH sales rep, or by calling 1-800-526-0485. Create an on-line outline of the content you want, in the order you want, using ESource's simple interface. Either type or cut and paste your own material and insert it into the text flow. You can preview the overall organization of the text you've created at any time. (Note, however, that since this preview is immediate, it comes unformatted.) Then, simply press another button and receive an order number for your own custom book. If you are not ready to order, do nothing—ESource will save your work. You can come back at any time and change, rearrange, or add more material to your creation. You are in control. Once you're finished and you have an ISBN, give it to your bookstore and your book will arrive on their shelves six weeks after they order. Your custom desk copies with their instructor supplements will arrive at your address at the same time.

To learn more about this new system for creating the perfect textbook, go to www.prenhall.com/esource. You can either go through the on-line walk-through of how to create a book, or experiment yourself.

Supplements

Adopters of ESource receive an instructor's CD that contains professor and student code from the books in the series, as well as other instruction aides provided by authors. The website also holds approximately **350 Powerpoint transparencies** created by Jack Leifer of University of Kentucky-Paducah available to download. Professors can either follow these transparencies as already prepared lectures or use them as the basis for their own custom presentations.

Titles in the ESource Series

Design Concepts for Engineers, 2/e
0-13-093430-5
Mark Horenstein

Engineering Success, 2/e
0-13-041827-7
Peter Schiavone

Engineering Design and Problem Solving, 2E
ISBN 0-13-093399-6
Steven K. Howell

Exploring Engineering
ISBN 0-13-093442-9
Joe King

Engineering Ethics
0-13-784224-4
Charles B. Fleddermann

Engineering Design—A Day in the Life of Four Engineers
0-13-085089-6
Mark N. Horenstein

Introduction to Engineering Analysis
0-13-016733-9
Kirk D. Hagen

Introduction to Engineering Experimentation
0-13-032835-9
Ronald W. Larsen, John T. Sears, and Royce Wilkinson

Introduction to Mechanical Engineering
0-13-019640-1
Robert Rizza

Introduction to Electrical and Computer Engineering
0-13-033363-8
Charles B. Fleddermann and Martin Bradshaw

Introduction to MATLAB 6
0-13-032845-6
Delores Etter and David C. Kuncicky, with Douglas W. Hull

Introduction to MATLAB
0-13-013149-0
Delores Etter with David C. Kuncicky

Introduction to Mathcad 2000
0-13-020007-7
Ronald W. Larsen

Introduction to Mathcad
0-13-937493-0
Ronald W. Larsen

Introduction to Mathcad 11
0-13-008177-9
Ronald W. Larsen

Introduction to Maple 8
0-13-032844-8
David I. Schwartz

Introduction to Maple
0-13-095133-1
David I. Schwartz

Mathematics Review
0-13-011501-0
Peter Schiavone

Power Programming with VBA/Excel
0-13-047377-4
Steven C. Chapra

http://emissary.prenhall.com/esource/

Introduction to Excel 2002
0-13-008175-2
David C. Kuncicky

Introduction to Excel, 2/e
0-13-016881-5
David C. Kuncicky

Engineering with Excel
ISBN 0-13-017696-6
Ronald W. Larsen

Introduction to Word 2002
0-13-008170-1
David C. Kuncicky

Introduction to Word
0-13-254764-3
David C. Kuncicky

Introduction to PowerPoint 2002
0-13-008179-5
Jack Leifer

Introduction to PowerPoint
0-13-040214-1
Jack Leifer

Graphics Concepts
0-13-030687-8
Richard M. Lueptow

Graphics Concepts with SolidWorks
0-13-014155-0
Richard M. Lueptow and Michael Minbiole

Graphics Concepts with Pro/ENGINEER
0-13-014154-2
Richard M. Lueptow, Jim Steger, and Michael T. Snyder

Introduction to AutoCAD 2000
0-13-016732-0
Mark Dix and Paul Riley

Introduction to AutoCAD, R. 14
0-13-011001-9
Mark Dix and Paul Riley

Introduction to UNIX
0-13-095135-8
David I. Schwartz

Introduction to the Internet, 3/e
0-13-031355-6
Scott D. James

Introduction to Visual Basic 6.0
0-13-026813-5
David I. Schneider

Introduction to C
0-13-011854-0
Delores Etter

Introduction to C++
0-13-011855-9
Delores Etter

Introduction to FORTRAN 90
0-13-013146-6
Larry Nyhoff and Sanford Leestma

Introduction to Java
0-13-919416-9
Stephen J. Chapman

About the Authors

No project could ever come to pass without a group of authors who have the vision and the courage to turn a stack of blank paper into a book. The authors in this series worked diligently to produce their books and provide the building blocks of the series.

Delores M. Etter is a Professor of Electrical and Computer Engineering at the University of Colorado. Dr. Etter was a faculty member at the University of New Mexico and also a Visiting Professor at Stanford University. Dr. Etter was responsible for the Freshman Engineering Program at the University of New Mexico and is active in the Integrated Teaching Laboratory at the University of Colorado. She was elected a Fellow of the Institute of Electrical and Electronics Engineers for her contributions to education and for her technical leadership in digital signal processing. In addition to writing best-selling textbooks for engineering computing, Dr. Etter has also published research in the area of adaptive signal processing.

Sanford Leestma is a Professor of Mathematics and Computer Science at Calvin College, and received his Ph.D. from New Mexico State University. He has been the longtime coauthor of successful textbooks on Fortran, Pascal, and data structures in Pascal. His current research interests are in the areas of algorithms and numerical computation.

Larry Nyhoff is a Professor of Mathematics and Computer Science at Calvin College. After doing bachelor's work at Calvin, and Master's work at Michigan, he received a Ph.D. from Michigan State and also did graduate work in computer science at Western Michigan. Dr. Nyhoff has taught at Calvin for the past 34 years—mathematics at first and computer science for the past several years. He has co-authored several computer science textbooks since 1981, including titles on Fortran and C++, as well as a brand new title on Data Structures in C++.

Acknowledgments: We express our sincere appreciation to all who helped in the preparation of this module, especially our acquisitions editor Alan Apt, managing editor Laura Steele, developmental editor Sandra Chavez, and production editor Judy Winthrop. We also thank Larry Genalo for several examples and exercises and Erin Fulp for the Internet address application in Chapter 10. We appreciate the insightful review provided by Bart Childs. We thank our families—Shar, Jeff, Dawn, Rebecca, Megan, Sara, Greg, Julie, Joshua, Derek, Tom, Joan, Marge, Michelle, Sandy, Lory, and Michael—for being patient and understanding. We thank God for allowing us to write this text.

Mark Dix began working with AutoCAD in 1985 as a programmer for CAD Support Associates, Inc. He helped design a system for creating estimates and bills of material directly from AutoCAD drawing databases for use in the automated conveyor industry. This system became the basis for systems still widely in use today. In 1986 he began collaborating with Paul Riley to create AutoCAD training materials, combining Riley's background in industrial design and training with Dix's background in writing, curriculum development, and programming. Dix and Riley have created tutorial and teaching methods for every AutoCAD release since Version 2.5. Mr. Dix has a Master of Education from the University of Massachusetts. He is currently the Director of Dearborn Academy High School in Arlington, Massachusetts.

Paul Riley is an author, instructor, and designer specializing in graphics and design for multimedia. He is a founding partner of CAD Support Associates, a contract service and professional training organization for computer-aided design. His 15 years of business experience and 20 years of teaching experience are supported by degrees in education and computer science. Paul has taught AutoCAD at the University of Massachusetts at Lowell and is presently teaching AutoCAD at Mt. Ida College in Newton, Massachusetts. He has developed a program, Computer-aided Design for Professionals, that is highly regarded by corporate clients and has been an ongoing success since 1982.

Scott D. James is a staff lecturer at Kettering University (formerly GMI Engineering & Management Institute) in Flint, Michigan. He is currently pursuing a Ph.D. in Systems Engineering with an emphasis on software engineering and computer-integrated manufacturing. Scott decided to write textbooks after he found a void in the books that were available. "I really wanted a book that showed how to do things in good detail, but in a clear and concise way. Many of the books on the market are full of fluff and force you to dig out the really important facts." Scott decided on teaching as a profession after several years in the computer industry. "I thought that it was really important to know what it was like outside of academia. I wanted to provide students with classes that were up to date and provide the information that is really used and needed."

Acknowledgments: Scott would like to acknowledge his family for allowing him the time to work on the text and his students and peers at Kettering who offered helpful critiques of the materials that eventually became the book.

Charles B. Fleddermann is a professor in the Department of Electrical and Computer Engineering at the University of New Mexico in Albuquerque, New Mexico. All of his degrees are in electrical engineering: his Bachelor's degree from the University of Notre Dame, and his Master's and Ph.D. from the University of Illinois at Urbana-Champaign. Prof. Fleddermann developed an engineering ethics course for his department in response to the ABET requirement to incorporate ethics topics into the undergraduate engineering curriculum. *Engineering Ethics* was written as a vehicle for presenting ethical theory, analysis, and problem solving to engineering undergraduates in a concise and readily accessible way.

Acknowledgments: I would like to thank Profs. Charles Harris and Michael Rabins of Texas A & M University, whose NSF sponsored workshops on engineering ethics got me started thinking in this field. Special thanks to my wife Liz, who proofread the manuscript for this book, provided many useful suggestions, and who helped me learn how to teach "soft" topics to engineers.

David I. Schwartz is an Assistant Professor in the Computer Science Department at Cornell University and earned his B.S., M.S., and Ph.D. degrees in Civil Engineering from State University of New York at Buffalo. Throughout his graduate studies, Schwartz combined principles of computer science with applications in civil engineering. He became interested in helping students learn how to apply software tools for solving a variety of engineering problems. He teaches his students to learn incrementally and practice frequently to gain the maturity to tackle other subjects. In his spare time, Schwartz plays drums in a variety of bands.

Acknowledgments: I dedicate my books to my family, friends, and students who all helped in so many ways. Many thanks go to the schools of Civil Engineering and Engineering & Applied Science at State University of New York at Buffalo where I originally developed and tested my UNIX and Maple books. I greatly appreciate the opportunity to explore my goals and all the help from everyone at the Computer Science Department at Cornell. Eric Svendsen and everyone at Prentice Hall also deserve my gratitude for helping to make these books a reality. Many thanks, also, to those who submitted interviews and images.

Ron Larsen is a professor in the Department of Chemical Engineering at Montana State University and received his Ph.D. from the Pennsylvania State University. He was initially attracted to engineering by the challenges the profession offers. Some of the greatest challenges he has faced while teaching have involved nontraditional teaching situations, including evening courses for practicing engineers and teaching through an interpreter at the Mongolian National University. These experiences provided opportunities to try new ways of communicating technical material, and the new techniques have been incorporated in his lectures and texts. *Introduction to Mathcad* and *Engineering with Excel* developed from sets of course notes prepared to help his students more effectively utilize a new software tool. They continue to be updated to meet the needs of engineers in an ever-changing workplace.

Acknowledgments: To my students at Montana State University, who have endured the rough drafts and the typos, and who still allow me to experiment in their classes; and to Pris, Anna, and Ben, who allow me to take some of their time to work on texts—my sincere thanks.

Peter Schiavone is a professor and student advisor in the Department of Mechanical Engineering at the University of Alberta, Canada. He received his Ph.D. from the University of Strathclyde, U.K. in 1988. He has authored several books in the area of student academic success, as well as numerous papers in international scientific research journals. Dr. Schiavone has worked in private industry in several different areas of engineering, including aerospace and systems engineering. He founded the first Mathematics Resource Center at the University of Alberta, a unit designed specifically to teach new students the necessary *survival skills* in mathematics and the physical sciences required for success in first-year engineering. This led to the Students' Union Gold Key Award for outstanding contributions to the university. Dr. Schiavone lectures regularly to freshman engineering students and to new engineering professors on engineering success, in particular about maximizing students' academic performance. He wrote the book *Engineering Success* in order to share the *secrets of success in engineering study*—the most effective, tried and tested methods used by the most successful engineering students.

Acknowledgements: Thanks to Eric Svendsen for his encouragement and support; to Richard Felder for being such an inspiration; to my wife Linda for sharing my dreams and believing in me; and to Francesca and Antonio for putting up with Dad when I was working on the text.

Mark N. Horenstein is a Professor in the Department of Electrical and Computer Engineering at Boston University. He has degrees in Electrical Engineering from M.I.T. and U.C. Berkeley and has been involved in teaching engineering design for the greater part of his academic career. He devised and developed the senior design project class taken by all electrical and computer engineering students at Boston University. In this class, the students work for a virtual engineering company developing products and systems for real-world engineering and social-service clients. Many of the design projects developed in his class have been aimed at assistive technologies for individuals with disabilities.

Acknowledgments: I would like to thank Prof. James Bethune, the architect of the Peak Performance event at Boston University, for his permission to highlight the competition in my text. Several of the ideas relating to brainstorming and teamwork were derived from a workshop on engineering design offered by Prof. Charles Lovas of Southern Methodist University. The principles of estimation were derived in part from a freshman engineering problem posed by Prof. Thomas Kincaid of Boston University.

Kirk D. Hagen is a professor at Weber State University in Ogden, Utah. He has taught introductory-level engineering courses and upper-division thermal science courses at WSU since 1993. He received his B.S. degree in physics from Weber State College and his M.S. degree in mechanical engineering from Utah State University, after which he worked as a thermal designer/analyst in the aerospace and electronics industries. After several years of engineering practice, he resumed his formal education, earning his Ph.D. in mechanical engineering at the University of Utah. Hagen is the author of an undergraduate heat transfer text. Having drawn upon his industrial and teaching experience, he strongly believes that engineering students must develop effective analytical problem-solving abilities. His book, *Introduction to Engineering Analysis*, was written to help beginning engineering students learn a systematic approach to engineering analysis.

Richard M. Lueptow is the Charles Deering McCormick Professor of Teaching Excellence and Associate Professor of Mechanical Engineering at Northwestern University. He is a native of Wisconsin and received his doctorate from the Massachusetts Institute of Technology in 1986. He teaches design, fluid mechanics, and spectral analysis techniques. "In my design class I saw a need for a self-paced tutorial for my students to learn CAD software quickly and easily. I worked with several students a few years ago to develop just this type of tutorial, which has since evolved into a book. My goal is to introduce students to engineering graphics and CAD, while showing them how much fun it can be." Rich has an active research program on rotating filtration, Taylor Couette flow, granular flow, fire suppression, and acoustics. He has five patents and over 40 refereed journal and proceedings papers along with many other articles, abstracts, and presentations.

Acknowledgments: Thanks to my talented and hardworking co-authors as well as the many colleagues and students who took the tutorial for a "test drive." Special thanks to Mike Minbiole for his major contributions to Graphics Concepts with SolidWorks. Thanks also to Northwestern University for the time to work on a book. Most of all, thanks to my loving wife, Maiya, and my children, Hannah and Kyle, for supporting me in this endeavor. (Photo courtesy of Evanston Photographic Studios, Inc.)

Jack Leifer is an Assistant Professor in the Department of Mechanical Engineering at the University of Kentucky Extended Campus Program in Paducah. He was previously with the Department of Mathematical Sciences and Engineering at the University of South Carolina–Aiken. He received his Ph.D. in Mechanical Engineering from the University of Texas at Austin in December 1995. His current research interests include the modeling of sensors for manufacturing, and the use of Artificial Neural Networks to predict corrosion.

Acknowledgements: I'd like to thank my colleagues at USC—Aiken, especially Professors Mike May and Laurene Fausett, for their encouragement and feedback; Eric Svendsen and Joe Russo of Prentice Hall, for their useful suggestions and flexibility with deadlines; and my parents, Felice and Morton Leifer, for being there and providing support (as always) as I completed this book.

Martin D. Bradshaw was born in Pittsburg, KS in 1936, grew up in Kansas and the surrounding states of Arkansas and Missouri, graduating from Newton High School, Newton, KS in 1954. He received the B.S.E.E. and M.S.E.E. degrees from the University of Wichita in 1958 and 1961, respectively. A Ford Foundation fellowship at Carnegie Institute of Technology followed from 1961 to 1963 and he received the Ph.D. degree in electrical engineering in 1964. He spent his entire academic career with the Department of Electrical and Computer Engineering at the University of New Mexico (1961-1963 and 1991-1996). He served as the Assistant Dean for Special Programs with the UNM College of Engineering from 1974 to 1976 and as the Associate Chairman for the EECE Department from 1993 to 1996. During the period 1987-1991 he was a consultant with his own company, EE Problem Solvers. During 1978 he spent a sabbatical year with the State Electricity Commission of Victoria, Melbourne, Australia. From 1979 to 1981 he served an IPA assignment as a Project Officer at the U.S. Air Force Weapons Laboratory, Kirkland AFB, Albuquerque, NM. He has won numerous local, regional, and national teaching awards, including the George Westinghouse Award from the ASEE in 1973. He was awarded the IEEE Centennial Medal in 2000.

Acknowledgments: Dr. Bradshaw would like to acknowledge his late mother, who gave him a great love of reading and learning, and his father, who taught him to persist until the job is finished. The encouragement of his wife, Jo, and his six children is a never-ending inspiration.

Stephen J. Chapman received a B.S. degree in Electrical Engineering from Louisiana State University (1975), the M.S.E. degree in Electrical Engineering from the University of Central Florida (1979), and pursued further graduate studies at Rice University. Mr. Chapman is currently Manager of Technical Systems for British Aerospace Australia, in Melbourne, Australia. In this position, he provides technical direction and design authority for the work of younger engineers within the company. He also continues to teach at local universities on a part-time basis.

Mr. Chapman is a Senior Member of the Institute of Electrical and Electronics Engineers (and several of its component societies). He is also a member of the Association for Computing Machinery and the Institution of Engineers (Australia).

**Steven C. Chapra** presently holds the Louis Berger Chair for Computing and Engineering in the Civil and Environmental Engineering Department at Tufts University. Dr. Chapra received engineering degrees from Manhattan College and the University of Michigan. Before joining the faculty at Tufts, he taught at Texas A&M University, the University of Colorado, and Imperial College, London. His research interests focus on surface water-quality modeling and advanced computer applications in environmental engineering. He has published over 50 refereed journal articles, 20 software packages and 6 books. He has received a number of awards including the 1987 ASEE Merriam/Wiley Distinguished Author Award, the 1993 Rudolph Hering Medal, and teaching awards from Texas A&M, the University of Colorado, and the Association of Environmental Engineering and Science Professors.

Acknowledgments: To the Berger Family for their many contributions to engineering education. I would also like to thank David Clough for his friendship and insights, John Walkenbach for his wonderful books, and my colleague Lee Minardi and my students Kenny William, Robert Viesca and Jennifer Edelmann for their suggestions.

**Robert Rizza** is an Assistant Professor of Mechanical Engineering at North Dakota State University, where he teaches courses in mechanics and computer-aided design. A native of Chicago, he received the Ph.D. degree from the Illinois Institute of Technology. He is also the author of *Getting Started with Pro/ENGINEER*. Dr. Rizza has worked on a diverse range of engineering projects including projects from the railroad, bioengineering, and aerospace industries. His current research interests include the fracture of composite materials, repair of cracked aircraft components, and loosening of prostheses.

**Steven Howell** is the Chairman and a Professor of Mechanical Engineering at Lawrence Technological University. Prior to joining LTU in 2001, Dr. Howell led a knowledge-based engineering project for Visteon Automotive Systems and taught computer-aided design classes for Ford Motor Company engineers. Dr. Howell also has a total of 15 years experience as an engineering faculty member at Northern Arizona University, the University of the Pacific, and the University of Zimbabwe. While at Northern Arizona University, he helped develop and implement an award-winning interdisciplinary series of design courses simulating a corporate engineering-design environment.

**Douglas W. Hull** is a graduate student in the Department of Mechanical Engineering at Carnegie Mellon University in Pittsburgh, Pennsylvania. He is the author of *Mastering Mechanics I Using Matlab 5*, and contributed to *Mechanics of Materials* by Bedford and Liechti. His research in the Sensor Based Planning lab involves motion planning for hyper-redundant manipulators, also known as serpentine robots.

**David C. Kuncicky** is a native Floridian. He earned his Baccalaureate in psychology, Master's in computer science, and Ph.D. in computer science from Florida State University. He has served as a faculty member in the Department of Electrical Engineering at the FAMU–FSU College of Engineering and the Department of Computer Science at Florida State University. He has taught computer science and computer engineering courses for over 15 years. He has published research in the areas of intelligent hybrid systems and neural networks. He is currently the Director of Engineering at Bioreason, Inc. in Sante Fe, New Mexico.

Acknowledgments: Thanks to Steffie and Helen for putting up with my late nights and long weekends at the computer. Finally, thanks to Susan Bassett for having faith in my abilities, and for providing continued tutelage and support.

**David I. Schneider** holds an A.B. degree from Oberlin College and a Ph.D. degree in Mathematics from MIT. He has taught for 34 years, primarily at the University of Maryland. Dr. Schneider has authored 28 books, with one-half

of them computer programming books. He has developed three customized software packages that are supplied as supplements to over 55 mathematics textbooks. His involvement with computers dates back to 1962, when he programmed a special purpose computer at MIT's Lincoln Laboratory to correct errors in a communications system.

Royce Wilkinson received his undergraduate degree in chemistry from Rose-Hulman Institute of Technology in 1991 and the Ph.D. degree in chemistry from Montana State University in 1998 with research in natural product isolation from fungi. He currently resides in Bozeman, MT and is involved in HIV drug research. His research interests center on biological molecules and their interactions in the search for pharmaceutical advances.

John T. Sears received the Ph.D. degree from Princeton University. Currently, he is a Professor and the head of the Department of Chemical Engineering at Montana State University. After leaving Princeton he worked in research at Brookhaven National Laboratory and Esso Research and Engineering, until he took a position at West Virginia University. He came to MSU in 1982, where he has served as the Director of the College of Engineering Minority Program and Interim Director for BioFilm Engineering. Prof. Sears has written a book on air pollution and economic development, and over 45 articles in engineering and engineering education.

Michael T. Snyder is President of Internet startup company Appointments123.com. He is a native of Chicago, and he received his Bachelor of Science degree in Mechanical Engineering from the University of Notre Dame. Mike also graduated with honors from Northwestern University's Kellogg Graduate School of Management in 1999 with his Masters of Management degree. Before Appointments123.com, Mike was a mechanical engineer in new product development for Motorola Cellular and Acco Office Products. He has received four patents for his mechanical design work. "Pro/ENGINEER was an invaluable design tool for me, and I am glad to help students learn the basics of Pro/ENGINEER."

Acknowledgments: Thanks to Rich Lueptow and Jim Steger for inviting me to be a part of this great project. Of course, thanks to my wife Gretchen for her support in my various projects.

Jim Steger is currently Chief Technical Officer and cofounder of an Internet applications company. He raduated with a Bachelor of Science degree in Mechanical Engineering from Northwestern University. His prior work included mechanical engineering assignments at Motorola and Acco Brands. At Motorola, Jim worked on part design for two-way radios and was one of the lead mechanical engineers on a cellular phone product line. At Acco Brands, Jim was the sole engineer on numerous office product designs. His Worx stapler has won design awards in the United States and in Europe. Jim has been a Pro/ENGINEER user for over six years.

Acknowledgments: Many thanks to my co-authors, especially Rich Lueptow for his leadership on this project. I would also like to thank my family for their continuous support.

Joe King received the B.S. and M.S. degrees from the University of California at Davis. He is a Professor of Computer Engineering at the University of the Pacific, Stockton, CA, where he teaches courses in digital design, computer design, artificial intelligence, and computer networking. Since joining the UOP faculty, Professor King has spent yearlong sabbaticals teaching in Zimbabwe, Singapore, and Finland. A licensed engineer in the state of California, King's industrial experience includes major design projects with Lawrence Livermore National Laboratory, as well as independent consulting projects. Prof. King has had a number of books published with titles including *Matlab*, MathCAD, Exploring Engineering, and Engineering and Society.

Reviewers

ESource benefited from a wealth of reviewers who on the series from its initial idea stage to its completion. Reviewers read manuscripts and contributed insightful comments that helped the authors write great books. We would like to thank everyone who helped us with this project.

Concept Document

Naeem Abdurrahman *University of Texas, Austin*
Grant Baker *University of Alaska, Anchorage*
Betty Barr *University of Houston*
William Beckwith *Clemson University*
Ramzi Bualuan *University of Notre Dame*
Dale Calkins *University of Washington*
Arthur Clausing *University of Illinois at Urbana–Champaign*
John Glover *University of Houston*
A.S. Hodel *Auburn University*
Denise Jackson *University of Tennessee, Knoxville*
Kathleen Kitto *Western Washington University*
Terry Kohutek *Texas A&M University*
Larry Richards *University of Virginia*
Avi Singhal *Arizona State University*
Joseph Wujek *University of California, Berkeley*
Mandochehr Zoghi *University of Dayton*

Books

Naeem Abdurrahman *University of Texas, Austin*
Stephen Allan *Utah State University*
Anil Bajaj *Purdue University*
Grant Baker *University of Alaska–Anchorage*
William Beckwith *Clemson University*
Haym Benaroya *Rutgers University*
John Biddle *California State Polytechnic University*
Tom Bledsaw *ITT Technical Institute*
Fred Boadu *Duke University*
Tom Bryson *University of Missouri, Rolla*
Ramzi Bualuan *University of Notre Dame*
Dan Budny *Purdue University*
Betty Burr *University of Houston*
Dale Calkins *University of Washington*
Harish Cherukuri *University of North Carolina –Charlotte*
Arthur Clausing *University of Illinois*
Barry Crittendon *Virginia Polytechnic and State University*
James Devine *University of South Florida*
Ron Eaglin *University of Central Florida*
Dale Elifrits *University of Missouri, Rolla*
Patrick Fitzhorn *Colorado State University*
Susan Freeman *Northeastern University*
Frank Gerlitz *Washtenaw College*
Frank Gerlitz *Washtenaw Community College*
John Glover *University of Houston*
John Graham *University of North Carolina–Charlotte*
Ashish Gupta *SUNY at Buffalo*
Otto Gygax *Oregon State University*
Malcom Heimer *Florida International University*
Donald Herling *Oregon State University*
Thomas Hill *SUNY at Buffalo*
A.S. Hodel *Auburn University*
James N. Jensen *SUNY at Buffalo*
Vern Johnson *University of Arizona*
Autar Kaw *University of South Florida*
Kathleen Kitto *Western Washington University*
Kenneth Klika *University of Akron*
Terry L. Kohutek *Texas A&M University*
Melvin J. Maron *University of Louisville*
Robert Montgomery *Purdue University*
Mark Nagurka *Marquette University*
Romarathnam Narasimhan *University of Miami*
Soronadi Nnaji *Florida A&M University*
Sheila O'Connor *Wichita State University*
Michael Peshkin *Northwestern University*
Dr. John Ray *University of Memphis*
Larry Richards *University of Virginia*
Marc H. Richman *Brown University*
Randy Shih *Oregon Institute of Technology*
Avi Singhal *Arizona State University*
Tim Sykes *Houston Community College*
Neil R. Thompson *University of Waterloo*
Dr. Raman Menon Unnikrishnan *Rochester Institute of Technology*
Michael S. Wells *Tennessee Tech University*
Joseph Wujek *University of California, Berkeley*
Edward Young *University of South Carolina*
Garry Young *Oklahoma State University*
Mandochehr Zoghi *University of Dayton*

Contents

1   MATHCAD: THE ENGINEER'S SCRATCH PAD   1

- 1.1   Introduction to Mathcad   1
- 1.2   Mathcad as a Design Tool   3
- 1.3   Mathcad as a Mathematical Problem Solver   4
- 1.4   Mathcad as a Unit Converter   4
- 1.5   Mathcad for Presenting Results   5
- 1.6   Mathcad's Place in an Engineer's Tool Kit   5
- 1.7   Objectives of the Text   6
- 1.8   Conventions Used in the Text   6

2   MATHCAD FUNDAMENTALS   8

- 2.1   The Mathcad Workplace   9
- 2.2   Determining the Order of Solving Equations in Mathcad   10
- 2.3   Four Different Kinds of Equal Signs   11
  - 2.3.1   Assignment $(:=)$ 12
  - 2.3.2   Display the Value of a Variable or Result of a Calculation $(=)$ 12
  - 2.3.3   Symbolic Equality $(=)$ 12
  - 2.3.4   Global Assignment $(\equiv)$ 12
- 2.4   Entering an Equation   13
  - 2.4.1   Predefined Variables   13
  - 2.4.2   Entering Nondecimal Values   14
  - 2.4.3   Exponents   14
  - 2.4.4   Selecting Part of an Equation   14
  - 2.4.5   Text Subscripts and Index Subscripts   15
  - 2.4.6   Changing a Value or a Variable Name   16
  - 2.4.7   Changing an Operator   16
  - 2.4.8   Inserting a Minus Sign   16
  - 2.4.9   Highlighting a Region   17
  - 2.4.10   Changing the Way Operators Are Displayed   17
- 2.5   Working With Units   19
  - 2.5.1   Defining a New Unit   22
  - 2.5.2   Editing the Units on a Value or Result   22
  - 2.5.3   Limitations to Mathcad's Units Capabilities   23
- 2.6   Controlling How Results Are Displayed   24
  - 2.6.1   Controlling the Way Numbers Are Displayed   24
  - 2.6.2   Controlling the Way Matrices Are Displayed   25
  - 2.6.3   Controlling the Way Units Are Displayed   26
- 2.7   Entering and Editing Text   27
  - 2.7.1   Sizing and Moving a Text Region or an Equation Region   29
  - 2.7.2   Inserting Equations inside Text Regions   29

    2.8   A Simple Editing Session   29
                2.8.1   Statement of the Problem   30

3 MATHCAD FUNCTIONS   42

3.1   Mathcad Functions   43
3.2   Elementary Mathematics Functions and Operators   44
      3.2.1   Common Mathematical Operators   44
      3.2.2   Logarithm and Exponentiation Functions   45
      3.2.3   Using QuickPlots to Visualize Functions   46
      3.2.4   3-d QuickPlots   46
3.3   Trigonometric Functions   47
      3.3.1   Standard Trigonometric Functions   47
      3.3.2   Inverse Trigonometric Functions   48
      3.3.3   Hyperbolic Trigonometric Functions   49
      3.3.4   Inverse Hyperbolic Trigonometric Functions   49
3.4   Advanced Mathematics Functions   50
      3.4.1   Round-off and Truncation Functions   50
      3.4.2   The if( ) Function   51
      3.4.3   Discontinuous Functions   52
      3.4.4   Boolean (Logical) Operators   52
      3.4.5   Random-Number-Generating Function   52
3.5   String Functions   53
3.6   User-Written Functions   53

4 WORKING WITH MATRICES   66

4.1   Mathcad's Matrix Definitions   67
4.2   Initializing an Array   68
4.3   Modifying an Array   80
4.4   Array Operations   87
4.5   Array Functions   94

5 DATA ANALYSIS FUNCTIONS   109

5.1   Graphing with Mathcad   110
5.2   Statistical Functions   117
5.3   Interpolation   118
5.4   Curve Fitting   122

6 PROGRAMMING IN MATHCAD   141

6.1   Mathcad Programs   142
6.2   Writing a Simple Program   143
6.3   The Programming Toolbar   144
      6.3.1   Add Line   145

	6.4	Program Flowcharts 145
	6.5	Basic Elements of Programming 148
		6.5.1 Data 148
		6.5.2 Input 152
		6.5.3 Operations 154
		6.5.4 Output 156
		6.5.5 Conditional Execution 161
		6.5.6 Loops 164
		6.5.7 Functions 169

7 MATHCAD'S SYMBOLIC MATH CAPABILITIES 181

7.1	Symbolic Math with Mathcad 182
7.2	Solving an Equation Symbolically 185
7.3	Manipulating Equations 186
	7.3.1 Substitution Using the Symbolic Keyword Toolbar 188
	7.3.2 Substitution Using the Symbolics Menu 188
7.4	Polynomial Coefficients 191
7.5	Symbolic Matrix Math 192
7.6	Symbolic Integration 193
	7.6.1 Symbolic Integration Using the Indefinite Integral Operator 194
	7.6.2 Symbolic Integration with Multiple Variables 195
	7.6.3 Symbolic Integration using the Symbolics Menu 196
	7.6.4 Definite Integrals: Symbolic Evaluation with Variable Limits 197
	7.6.5 Definite Integrals: Symbolic Evaluation with Numeric Limits 197
	7.6.6 Definite Integrals: Symbolic Evaluation with Mixed Limits 198
	7.6.7 Mixed Limits with Multiple Integration Variables 198
	7.6.8 Definite Integrals: Numerical Evaluation 198
7.7	Symbolic Differentiation 201
	7.7.1 First Derivative with Respect to one Variable 201
	7.7.2 Higher Derivatives with Respect to a Single Variable 203
	7.7.3 Differentiation with Respect to Multiple Variables 203

8 NUMERICAL TECHNIQUES 216

8.1	Iterative Solutions 217
8.2	Numerical Integration 222
	8.2.1 Integration 222
	8.2.2 A Trapezoidal Rule Function 227
	8.2.3 Simpson's-Rule Integration 229
8.3	Numerical Differentiation 233
	8.3.1 Using a Fitting Function 234
	8.3.2 Using Numerical Approximations for Derivatives 234

INDEX 246

1
Mathcad: The Engineer's Scratch Pad

1.1 INTRODUCTION TO MATHCAD

Mathcad[1] is an equation-solving software package that has proven to have a wide range of applicability to engineering problems. Mathcad's ability to display equations the same way you would write them on paper makes a Mathcad worksheet easy to read. For example, if you wanted to calculate the mass of water in a storage tank, you might solve the problem on paper as follows:

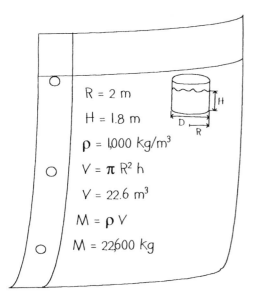

SECTIONS

1.1 Introduction to Mathcad
1.2 Mathcad as a Design Tool
1.3 Mathcad as a Mathematical Problem Solver
1.4 Mathcad as a Unit Converter
1.5 Mathcad for Presenting Results
1.6 Mathcad's Place in an Engineer's Tool Kit
1.7 Objectives of the Text
1.8 Conventions Used in the Text

OBJECTIVES

After reading this chapter, you will

- begin to see how Mathcad solves problems
- understand how Mathcad can assist the engineering design process
- see how Mathcad can be used to solve math problems
- learn how Mathcad handles unit conversions
- see how Mathcad can help you present your results to others
- know what to expect from the text
- become familiar with some nomenclature conventions that are used throughout the text

---

[1]Mathcad is a registered trademark of Mathsoft Engineering & Education, Inc., of Cambridge, Massachusetts.

The same calculation in Mathcad might look like this:

$$R := 2 \cdot m$$
$$H := 1.8 \cdot m$$
$$\rho := 1000 \cdot \frac{kg}{m^3}$$
$$V := \pi \cdot R^2 \cdot H$$
$$V = 22.6 \cdot m^3$$
$$M := \rho \cdot V$$
$$M = 2.26 \cdot 10^4 \cdot kg$$

One of the nicest features of Mathcad is its ability to solve problems much the same way people do, rather than making your solution process fit the program's way of doing things. For example, you could also solve this problem in a spreadsheet by entering the constants and the equations for volume and mass into various cells:

	A	B	C
1	R:	2	
2	H:	1.8	
3	Rho:	1000	
4			
5	V:	22.6	
6	M:	22619	
7			

In the C programming language, the problem might be solved with the following program:

```
#include stdio.h
#include math.h

main()
{
    float R, H, Rho, V, M;

    R = 2;
    H = 1.8;
    V = 3.1416 * pow (R, 2) * H;
    M = Rho * V;
    printf("V = %f \n M = % f", V, M);
}
```

While spreadsheets and programming languages can produce the solution, Mathcad's presentation is much more like the way people solve equations on paper. This makes Mathcad easier for you to use. It also makes it easier for others to read and understand your results.

PROFESSIONAL SUCCESS

Work to develop communication skills as well as technical skills.

An engineer's job is to find solutions to technical problems. Finding a solution requires good technical skills, but the solution must *always* be communicated to other people. An engineer's communication skills are just as important as her or his technical skills.

Because Mathcad's worksheets are easy to read, they can help you communicate your results to others.

You can improve the readability of your worksheets by

- performing your calculations in an orderly way (plan your work),
- adding comments to your worksheet, and
- using units on your variables.

The last two items will be discussed in more detail in the next chapter.

Mathcad's user interface is an important feature, but Mathcad has other features that make it excel as a design tool, a mathematical problem solver, a unit converter, and a communicator of results.

1.2 MATHCAD AS A DESIGN TOOL

A Mathcad *worksheet* is a collection of variable definitions, equations, text regions, and graphs displayed on the screen in pretty much the same fashion you would write them on paper. A big difference between a Mathcad worksheet and your paper scratch pad is *automatic recalculation*: If you make a change to any of the definitions or equations in your worksheet, the rest of the worksheet is automatically updated. This makes it easy to do the "what if" calculations that are so common in engineering. For example, what if the water level rises to 2.8 m? Would the mass in the tank exceed the tank's maximum design value of 40,000 kg?

To answer these questions, simply edit the definition of H in the Mathcad worksheet. The rest of the equations are automatically updated, and the new result is displayed:

$$
\begin{aligned}
R &:= 2 \cdot m \\
H &:= 2.8 \cdot m \\
\rho &:= 1000 \cdot \frac{kg}{m^3} \\
V &:= \pi \cdot R^2 \cdot H \\
V &:= 35.2 \cdot m^3 \\
M &:= \rho \cdot V \\
M &= 3.52 \cdot 10^4 \cdot kg
\end{aligned}
$$

From this calculation, we see that even at a height of 2.8 m, the mass in the tank is still within the design specifications.

The ability to develop a worksheet for a particular case and then vary one or more parameters to observe their impact on the calculated results makes a Mathcad worksheet a valuable tool for evaluating multiple designs.

Mathcad has another feature, called a QuickPlot, that is very useful for visualizing functions, and this can also speed the design process. With a QuickPlot, you simply create a graph, put the function on the *y*-axis, and place a dummy variable on the *x*-axis. Mathcad evaluates the function for a range of values and displays the graph.

For example, if you want to see what the hyperbolic sine function looks like, use a QuickPlot, such as the following:

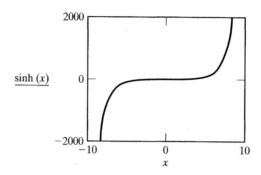

QuickPlots are discussed in more detail in later chapters—we will use them a lot.

1.3 MATHCAD AS A MATHEMATICAL PROBLEM SOLVER

Mathcad has the ability to solve problems numerically (computing a value) or symbolically (working with the variables directly). It has a large collection of built-in functions for trigonometric calculations, statistical calculations, data analysis (e.g., regression), and matrix operations. Mathcad can calculate derivatives and evaluate integrals, and it can handle many differential equations. It can work with imaginary numbers and handle Laplace transforms. Iterative solutions are tedious by hand, but very straightforward in Mathcad. In sum, while it is possible to come up with problems that are beyond Mathcad's capabilities, Mathcad can handle the bulk of an engineer's day-to-day calculations—and do them very well.

Some of these features, like Laplace transforms and functions for solving differential equations, are beyond the scope of this text, but many of Mathcad's commonly used features and functions will be presented and used.

1.4 MATHCAD AS A UNIT CONVERTER

Mathcad allows you to build units into most equations. (There are a couple of restrictions, which will be discussed in the next chapter.) Allowing Mathcad to handle the chore of getting all the values converted to a consistent set of units can be a major timesaver. For example, if the storage tank's radius had been measured in feet and the depth in inches, then

$$R = 6.56 \text{ ft (equivalent to 2 m)}$$

$$H = 70.87 \text{ in (equivalent to 1.8 m)}$$

We could convert the feet and inches to meters ourselves; or we can let Mathcad do the job, as shown here:

$$R := 6.56 \cdot \text{ft}$$
$$H := 70.87 \cdot \text{in}$$
$$\rho := 1000 \cdot \frac{\text{kg}}{\text{m}^3}$$
$$V := \pi \cdot R^2 \cdot H$$
$$V := 22.609 \cdot \text{m}^3$$

$$M := \rho \cdot V$$
$$M = 22609 \cdot kg$$

Note: The mass is slightly lower here than in the first example because of rounding of the R and H values.

For complicated problems, and when your input values have many different units, Mathcad's ability to handle the unit conversions is a very nice feature.

1.5 MATHCAD FOR PRESENTING RESULTS

Engineering and science are both fields in which a person's computed results have little meaning unless they are given to someone else. A circuit design has to be passed along to a manufacturer, for example. Getting the results in a useful form can often take as long as computing the results. This is an area where Mathcad can help speed up the process.

Practicing engineers use spreadsheets for many routine calculations, but spreadsheets can be frustrating because they display only the calculated results while hiding the equations. If someone gives you a spreadsheet printout, it probably shows only numbers, and you have to take the person's word for the equations or ask for a copy of the spreadsheet file. If you dig into the spreadsheet to see the equations, they are still somewhat cryptic because they typically use cell references (e.g., B12) rather than variable names.

Computer program listings make it clear how the results were calculated, and recognizable variable names can be used, but long equations on a single line can still be difficult to decipher. Mathcad's ability to show the equations and the results the way people are used to reading them makes a Mathcad worksheet a good way to give your results to someone else. If there is a question about a result, the solution method is obvious. Mathcad's ability to print equations in the order we would write them, but with typeset quality, is a bonus.

Still, there are times when you need to get your results into a more formal report. Mathcad helps out in that area too. Equations and results (e.g., values, matrices, graphs) on a Mathcad worksheet can be inserted (via copy and paste operations) into other software programs, such as word processors. You don't have to retype the equations in the word processor, which can be a major time-saver. This compatibility with other software also means you can insert a matrix from Mathcad into a spreadsheet or create a Mathcad matrix from a column of values in a spreadsheet. These features will be described in much more detail later.

1.6 MATHCAD'S PLACE IN AN ENGINEER'S TOOL KIT

Spreadsheets, programming languages, and mathematical problem solvers all have their place, and the tools you will use routinely will depend on where your career takes you, or perhaps vice versa. There is no single "right" tool for most problems, but Mathcad seems a logical choice when the requirements of the problem align with Mathcad's strengths, which include

- equations that are displayed in a highly readable form,
- the ability to work with units,
- a symbolic math capability,
- an iterative solution capability, and
- an extensive function library.

Deciding on which software product to use requires an understanding of the various products. For example,

- Spreadsheets can solve equations requiring iteration, but the process is much easier to follow in Mathcad.
- Mathcad can handle lists of numbers (e.g., analyses of experimental data sets), but columns of numbers fit well into the strengths of a spreadsheet.
- Spreadsheets cannot handle symbolic mathematics, so Mathcad (or Maple, or Mathematica)[2] must be used for that type of work.

This text demonstrates some of Mathcad's capabilities; it is my hope that it also will help you learn where the package fits in your tool kit.

1.7 OBJECTIVES OF THE TEXT

The first objective of the text is to teach you how to use Mathcad to solve engineering problems. The second objective is to show the wide range of career areas open to engineers and to demonstrate how Mathcad fits into all these areas. Finally, the third objective is to address a few of the challenges and opportunities that the next generation of engineers will face. With these objectives in mind, a typical chapter includes the following features:

- an introduction to one of the expanding fields of engineering;
- information on Mathcad, initially using simple examples for clarity;
- "Practice!" boxes—an opportunity for you to try Mathcad's features for yourself, using quick and easy problems;
- application boxes, showing how Mathcad can be used to solve real problems.

At the end of each chapter are several homework problems, some related to the challenges and opportunities mentioned. The following themes appear throughout the text:

- *nanotechnology*, a whole new area of opportunities for the next generation of engineers;
- *biomedical engineering*, an opportunity for future engineers;
- *optics*, an old field that is expanding rapidly into new areas;
- *risk analysis*, a necessary and challenging part of many engineering activities;
- *total recycle*, seeking to eliminate waste products entirely;
- *composite materials*, engineered materials that require the skills of a variety of engineering disciplines and that have the potential to change the way we do a lot of things;
- *alternative energy*; meeting the world's energy needs in the future.

---

[2]Maple is a product of Waterloo Maple, Inc., 57 Erb Street, W. Waterloo, Ontario, Canada. Mathematica is produced by Wolfram Research, Inc., Champaign, Illinois.

1.8 CONVENTIONS USED IN THE TEXT

`Stdev(v)`	Function names and variable names in text are shown in a different font.
[Ctrl-6]	Keystrokes are shown in brackets. This example indicates that the control key and 6 key should be pressed simultaneously.
File/Save As	Menu selections are listed with the main menu item and sub-menu items separated by "/".
`A := `$\pi \cdot r^2$	Mathcad examples are shown in the Courier font and are indented.
Key term	Key terms are shown in italics the first time they are used.

KEY TERMS

automatic recalculation
design

unit conversion

worksheet

2

Mathcad Fundamentals

NANOTECHNOLOGY

The field of *nanotechnology* is relatively new, but growing rapidly. In general terms, it is the science of the very small. (One nanometer is 1×10^{-9} meter.) The goal of this new field of study is to develop the ability to manufacture on a molecular level, arranging individual atoms to create structures and devices with particular characteristics, properties, or abilities. The types of devices envisioned sound a lot like science fiction:

- Computer chips much smaller, faster, and vastly more powerful than any available today.
- Molecular-sized machines capable of detecting and destroying disease-causing microorganisms.
- Devices that can recognize toxic molecules and rearrange them, making them innocuous.
- Building and road surfaces designed to capture and utilize solar energy.

Developers in this area have to be able to think outside the box because developing molecular-sized devices requires new approaches and a reexamination of all of the assumptions associated with macro-scale designs. But the individuals who work in this area will also need to be well grounded in the basic principles of science, since these are the rules that will govern the nanoscale design process. This

SECTIONS

2.1 The Mathcad Workplace
2.2 Determining the Order of Solving Equations in Mathcad
2.3 Four Different Kinds of Equal Signs!
2.4 Entering an Equation
2.5 Working with Units
2.6 Controlling How Results Are Displayed
2.7 Entering and Editing Text
2.8 A Simple Editing Session

OBJECTIVES

After reading this chapter, you will

- understand how the Mathcad workspace is laid out
- know how to enter an equation
- know how to display results and control how they are displayed
- understand how Mathcad determined the order in which to evaluate equations
- know how Mathcad handles unit conversions
- be able to enter and format text regions on a Mathcad worksheet

is also an area that will require cooperation and teamwork, since a mechanical engineer's knowledge of how to design a full-sized device must be combined with a knowledge of molecular behavior (chemists, biochemists, and chemical engineers) and knowledge of material and electrical properties of matter (material scientists, physicists, and electrical engineers). Finally, information on biological techniques from microbiologists and molecular biologists will be needed to build the devices.

Writers of science fiction have always attempted to predict the future by extrapolating the capabilities of today's science. While it is likely that nanotechnology will take some different paths than those being discussed today, it is a field that will certainly impact our lives in some ways in the near future.

PROFESSIONAL SUCCESS

Practice, Practice, Practice!
This does not mean that you should perform the same tasks over and over again (you're not trying to improve your manual dexterity), but you can take conceptual information in a textbook and make it your own by putting it into practice.

The fastest way to learn Mathcad is to use it. Try working through the examples in this text on a computer. There are "Practice!" boxes throughout the text that have been designed to help you learn what Mathcad can do—to help you put this new knowledge into practice.

Solutions to the Practice! problems are available at the text's companion website:

http://www.coe.montana.edu/che/mathcad

2.1 THE MATHCAD WORKPLACE

When you start Mathcad, you should see the following on your screen:

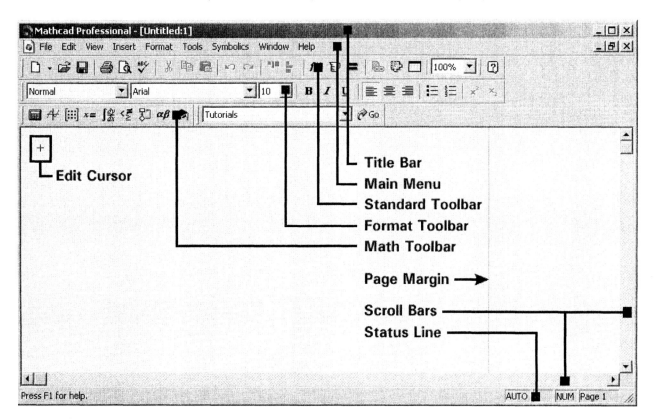

(The exact appearance of the screen depends upon the version of Mathcad you are using and the options that have been selected. Some of these options will be described in this chapter. The screen examples in this text are from Mathcad 11.)

Several of the features near the top of the screen should be pretty familiar to Windows users. The *Title Bar* and *Menu Bar* are common to most Windows applications. Many of the most common menu commands are also available as buttons on the Toolbar. The *Format Toolbar* is very similar to that used by most word processors, but in Mathcad it displays the font type and size of the *style* applied to the item you are editing.

There are separate styles defined for constants, variables, and text. You can change the formatting by using the Format Toolbar. When you change the format of *any* constant or variable, you change the format of *all* constants or variables in the worksheet. However, in text regions, only the selected portion of text is changed when you use the Format Toolbar. To change the way all text regions are displayed, modify the Normal text style, using the Format menu.

The *Math Toolbar*, shown just under the Format Bar toward the top of the screen (but it can be moved to other locations), is unique to Mathcad and provides access to a variety of useful mathematical symbols and functions. Clicking on any of the buttons on the Math Toolbar causes another toolbar to be displayed. For example, clicking the *Matrix Toolbar* button displays the Matrix Toolbar—a collection of functions that are useful for performing matrix operations. The Mathcad Toolbars available from the Math Toolbar include the following:

- *Calculator Toolbar*
- *Evaluation and Boolean Toolbar*
- *Graph Toolbar*
- *Vector and Matrix Toolbar*
- *Calculus Toolbar*
- *Programming Toolbar*
- *Greek Symbol Toolbar*
- *Symbolic Keyword Toolbar*

The majority of the workspace is a blank, white space called the *worksheet*. This is the area available for you to enter your equations, text, graphs, etc. The worksheet scrolls if you need more space.

There is a small crosshair cursor displayed on the worksheet. This is the *edit cursor*, and it indicates where the next equation or text region will be displayed. Clicking the mouse anywhere on the worksheet moves the edit cursor to the mouse pointer's location. If the edit cursor is located between two equations, you can add lines between the equations by pressing [Enter] or delete lines between the equations by pressing [Delete].

2.2 DETERMINING THE ORDER OF SOLVING EQUATIONS IN MATHCAD

It is usually very important to solve a set of equations in a particular order. In Mathcad, you use the placement of equations on the worksheet to control the order of their solution. Mathcad evaluates equations from left to right and top to bottom. When two equations are side by side on the worksheet, the equation on the left will be evaluated first, followed by the equation on the right. When there are no more equations to evaluate on a line, Mathcad moves down the worksheet and continues evaluating equations from left to right.

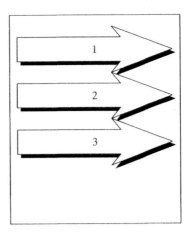

Some equations take a lot more space on the screen than others, so how do you determine which equation will be evaluated first? Mathcad assigns each equation an *anchor point* on the screen. The anchor point is located to the left of the first character in the equation, at the character baseline. You can ask Mathcad to display the anchor points by selecting Regions from the View Menu (View / Regions). When View / Regions is on, the background of the worksheet is dimmed, and the equation regions appear as bright boxes. The anchor point is indicated by a black dot.

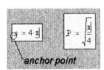

In the preceding figure, the equation on the left would be evaluated first, since the anchor points are the same distance from the top of the worksheet, but the anchor point on the s is to the left of the anchor point on the P. For simple variable definitions, like those of s and P shown here, the anchor point is always located at the bottom-left corner of the variable name.

2.3 FOUR DIFFERENT KINDS OF EQUAL SIGNS!

In high school algebra, you learned that the equal sign indicates that the left and right sides of an equation are equal. Then, in a programming course, you may have seen a statement such as

```
COUNT = COUNT + 1
```

This statement is never an algebraic equality, but it is a valid programming statement because the equal sign, in a programming context, means *assignment*, not algebraic equality. That is, the result of the calculation on the right side is assigned to the variable on the left. Mathcad allows both types of usage (algebraic equality and assignment) within a single worksheet, but different types of equal signs are used to keep things straight.

In Mathcad, four different symbols are available to represent equality or assignment (more if you count the "symbolic evaluation" and the programming "temporary assignment" symbols). Fortunately, some are used less often than others.

2.3.1 Assignment (:=)

The most commonly used *assignment operator* is :=, which is entered by using the colon key [:]. It is often called the "define as equal to" operator. This type of equal sign was used to assign values to variables s and P in the previous figure.

2.3.2 Display the Value of a Variable or Result of a Calculation (=)

Once a variable has been assigned a value (either directly or by means of a calculation), you can display the value by using the plain equal sign [=]. Displaying a value is the only usage of the plain equal sign in Mathcad; it is never used for assignment or algebraic equality. To display the value of the calculated variable s, you first define the equation used to compute s and then display the result using the equal sign. For example, we might have

```
s := 4•π
s = 12.566
```

You can also display a calculated result without assigning it to a variable. This displays the result of the calculation, but you cannot use that result in further calculations without assigning it to a variable. An example is

```
4•π = 12.566
```

2.3.3 Symbolic Equality (=)

A *symbolic equality* is used to indicate that the combination of variables on the left side of an equation is equal to the combination of variables on the right side of an equation—the high school algebra meaning of the equal sign. Symbolic equality is shown as a heavy boldface equal sign in Mathcad and is entered by pressing [Ctrl =]. (Hold down the control key while pressing the equal key.) An example of a formula that uses this operation is

```
P•V = n•R•T
```

Symbolic equality is used to show a relationship between variables. There is no assignment of a value to any variable when symbolic equality is used. This type of equal sign is used for symbolic math (see Chapter 7) and for solving equations by means of iterative methods (See Section 8.1).

2.3.4 Global Assignment (≡)

There is one way to override the left-to-right, top-to-bottom evaluation order in Mathcad: use a *global assignment* operator. Mathcad actually evaluates a worksheet in two passes. In the first pass, all global assignment statements are evaluated (from left to right and top to bottom). Then, in the second pass, all other equations are evaluated. Defining a variable with a global assignment equal sign has the same effect as putting the equation at the top of the worksheet, since both cause the statement to be evaluated first.

Global assignments are not used a lot, but it is fairly common to use them for unit definitions. For example, Mathcad already knows what a year (yr) is, but you could define a new unit, decade, in terms of years, as

```
decade ≡ 10•yr
```

Note: You could also define the decade by using the regular assignment operator, :=. It is common, but not required, to use global assignment for units.

PRACTICE!

What will Mathcad display as the value of x in each of the examples that follow? (Try each in a separate worksheet.)

a. Use of "define as equal to" ...
   y := 3
   x := y
   x =

b. Use of a symbolic equality ...
   y = 3
   x := y
   x =

c. With units ...
   y := 3 · cm
   x := y
   x =

d. Use of a "global define as equal to" ...
   x := y
   x =
   y ≡ 3

2.4 ENTERING AN EQUATION

To enter an equation, you simply position the edit cursor (crosshair) where you want the equation to go, and start typing. Mathcad creates an *equation region* and displays the equation as you enter it. (Whenever Mathcad is waiting for you to type in the workspace, it waits in *equation edit mode* so that you can easily enter a new equation.) To enter the defining equation for the variable s, you would type [s] [:] [4] [*] [Ctrl-Shift-p][Enter].[1] To see the result of this calculation, move the cursor to the right or down (or both), and type [s] [=]. Mathcad will display the result after the equal sign:

   s := 4 · π
   s = 12.566

2.4.1 Predefined Variables

Pi is such a commonly used value that it comes as a *predefined variable* in Mathcad. You can get the π symbol either from the Greek Symbols Toolbar or by pressing [Ctrl-Shift-p], the shortcut used in the previous paragraph. Pi comes predefined in Mathcad with a value of 3.14159265.... You can redefine pi (or any other predefined variable) simply by building it into a new definition:

   π := 7
   s := 4 · π
   s = 28

However, redefining a commonly used constant is not a good idea in most situations.

Four common predefined values in Mathcad are π, e, g, and %. These are entered by using [Ctrl-Shift-p], [e], [g], and [%], respectively.

---

[1] The keyboard shortcut for pi is [Ctrl-Shift-p] in recent versions of Mathcad, but prior to version 8 the shortcut was [Ctrl-p].

2.4.2 Entering Nondecimal Values

To enter a hexadecimal value when defining a variable, simply add the letter h after the value. Hexadecimal values can contain the digits 0 through 9 and the letters a through f. No decimal points are allowed. Similarly, octal values can be used in variable definitions simply by adding the letter o after the value. Only digits 0 through 7 may be used. Mathcad allows binary values (consisting of zeroes and ones only) to be entered by adding a letter b at the end of the value. The following examples are representative:

```
A := 12        A = 12      decimal
B := 12o       B = 10      octal
C := 12h       C = 18      hexadecimal
D := 1011b     D = 11      binary
```

To see a result expressed as an octal, a hexadecimal, or a binary value, double-click the value and change the *radix* of the displayed result on the Result Format dialog on the Display Options panel:

```
E := 201       E = 201
               E = 311o
               E = 0c9h
               E = 11001001b
```

2.4.3 Exponents

There are two ways to enter an exponent in an equation: you can use the *caret* [^] (or [Shift-6]) or, in Mathcad 11, press the superscript button on the Format Toolbar. The right-hand side of the equation for the area of a circle would be entered as [Ctrl-Shift-p] [*] [r] [^] [2], which gives

$$\text{Area} := \pi \cdot r^2$$

Note that once Mathcad has moved the edit cursor up so that you can enter the exponent, it *stays* in the exponent. That is, if you were to enter a plus sign after the 2 in the preceding equation, you would be adding to the 2 (the exponent), not to the r^2. If you need to add something to the r^2, you must first select a portion of the equation and then enter the plus sign.

Note: Mathcad uses standard mathematical *operator precedence rules:* exponentiation before multiplication and division, and multiplication and division before addition and subtraction. Parentheses can be used to ensure that an equation is evaluated in the desired order.

2.4.4 Selecting Part of an Equation

If you want to compute the surface area of a cylinder, you need to add the areas of the circles on each end ($2\pi r^2$) and the area of the side of the cylinder ($2\pi rL$) After entering the exponent on the first term, you need to add another piece to the equation. If you don't select the r^2 before entering the plus sign, you will end up adding to the exponent as follows:

$$A_{cyl} := 2 \cdot \pi \cdot r^{2 + 2 \cdot \pi \cdot r}$$

This is obviously not what we want. To get what we want, right after entering the 2 in the r^2, you need to press the [Space] key once to select the r^2. (Mathcad will indicate the selected portion of the equation with an underline. There is also a vertical line, called the *insert bar*, that shows where the next typed character will go.) After you enter the exponent, the worksheet should look like this:

$$A_{cyl} := 2 \cdot \pi \cdot r^2|$$

After you enter the exponent and press [Space], the worksheet should look like this:

$$A_{cyl} := 2 \cdot \pi \cdot \underline{r^2}$$

The underline beneath the r^2 indicates that the next operation will be applied to the entire selected region, which is what we want—we want to add $2\pi r L$ to the selected region. The final result is

$$A_{cyl} := 2 \cdot \pi \cdot r^2 + 2 \cdot \pi \cdot r \cdot L$$

Note: You might think you should select the entire term, $2 \cdot \pi \cdot r^2$, instead of just the r^2. That would work, too. You would just press [Space] two more times to select the π and the 2. To reduce the number of keystrokes required to enter an equation, Mathcad keeps multiplied variables together when you add to the collection. This is just a convenience; it is handy, but does take some getting used to.

PRACTICE!

What does Mathcad display when you enter the following key sequences?

a. [P][*][V][Ctrl =][n] [*] [R] [*] [T]
b. [P][:][n] [*] [R] [*] [T][/][V]
c. [P][:][n] [*] [R] [*] [T][Space][Space] [/][V]
d. [r][:][k][1][*][C][A][^][2][–][k][2][*][C][B]
e. [r][:][k][1][*][C][A][^][2][Space][–][k][2][*][C][B]

2.4.5 Text Subscripts and Index Subscripts

The "cyl" in the variable name A_{cyl} presented earlier is slightly lower than the A—this is an example of a *text subscript*. Mathcad will allow the use of a text subscript as part of a variable name, which can be useful when one is naming related variables. For example, the areas of a circle, sphere, and cylinder might be indicated as A_{circle}, A_{sphere}, and A_{cyl}, respectively. The variable name A_{cyl} was entered as [A] [.] [c] [y] [l], where the period was used to indicate that a text subscript follows.

The *index subscript*, which may look similar to the text subscript, is used for an entirely different purpose. Index subscripts indicate a particular element of an array (a vector or matrix). The first element of an array is called element zero in Mathcad. For example, if you have a three-element array called Z, containing the values 2, 5, and 7—that is,

$$Z := \begin{bmatrix} 2 \\ 5 \\ 7 \end{bmatrix}$$

then the value of element zero of array Z is 2. The zero element can be accessed individually using an index subscript, as in

$$Z_0 = 2$$

While it looks similar to a text subscript, an index subscript is entered differently and has a different meaning than a text subscript. The preceding index subscript was entered as [Z] [[ ] [0], where [[ ] means the left-square-bracket key.

PRACTICE!

> What does Mathcad display when you enter the following key sequences?
> a. [r][.][A][:][k][.][0][*][C][.][A][^][2]—this expression has a *text* subscript on k
> b. [r][.][A][:][k][[ ][0][*][C][.][A][^][2]—this expression has an *index* subscript on k
> c. [r][.][A][:][k][[ ][0][space][*][C][.][A][^][2]

2.4.6 Changing a Value or a Variable Name

If you click in the middle of a value or a variable name, a vertical insert bar (cursor) will be displayed. The insert bar indicates where any edits will take place. You can use the left- and right-arrow keys to move the bar. The delete key will remove the character to the right of the insert bar, while the backspace key will remove the character to the left.

Not all characters that are entered are displayed by Mathcad, but they can still be deleted. For example, the [.] used to enter a text subscript is not shown on the screen, but if you position the insert bar at the beginning of the subscript text and press [Backspace], the [.] will be removed, and the subscript will become regular text.

For example, if the ideal-gas constant (0.08206 liter atm/mole K) had been entered incorrectly as 0.08506, you would click to the right of the numeral 5, as in the following figure:

$$R_{gas} := 0.085|06$$

Then press [Backspace] to remove the 5, and press [2] to enter the correct value. The final result would look like this:

$$R_{gas} := 0.08206$$

2.4.7 Changing an Operator

When you need to change an operator, such as a symbolic equality (=) to an assignment operator (:=), you click just to the right of the operator itself. This puts insert bars around the character just to the right of the operator. Press [Insert] to move the vertical bar to the left side of the character (just to the right of the operator):

$$s = |4 \cdot \pi$$

Then press [Backspace] to delete the operator. An open placeholder appears, indicating where the new operator will be placed:

$$s \; \square \; |4 \cdot \pi$$

Enter the new assignment operator by pressing the colon key [:]:

$$s := |4 \cdot \pi$$

2.4.8 Inserting a Minus Sign

Special care must be taken when you need to insert a minus sign, because the same key is used to indicate both negation and subtraction. Here is how Mathcad decides which symbol to insert when you press the [−] key: If the insert bar is to the left of a character (to the right of the open placeholder), as in

$$A := 12 \cdot x^2 \; \square \; |5 \cdot y^2$$

then the sign of the character to the right of the placeholder is changed (indicating negation):

$$A := 12 \cdot x^2 \;\square\; -5 \cdot y^2$$

If the insert bar is to the right of a character (to the left of the open placeholder), as in

$$A := 12 \cdot x^2 | \square 5 \cdot y^2$$

then the open placeholder is filled by a subtraction operator:

$$A := 12 \cdot x^2 - 5 \cdot y^2$$

2.4.9 Highlighting a Region

To make your results stand out, Mathcad allows you to show them in a *highlighted region*. The highlighting can consist of a border around the result or a colored background.

To highlight a result, right-click on it, and select Properties from the pop-up menu:

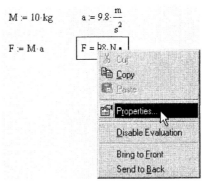

On the Properties dialog, select Highlight Region and Choose Color, or check the Show Border box:

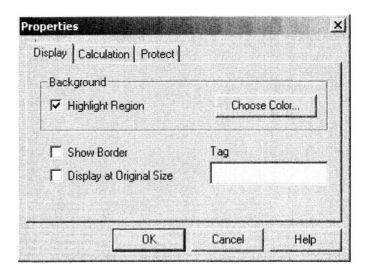

2.4.10 Changing the Way Operators Are Displayed

Mathcad uses specific symbols for operators, such as the := to indicate assignment. While these symbols help you read to a Mathcad worksheet, they can confuse people who are not used to working with the software. You can change the way operators are displayed using the Math Options dialog. Several of the common operators used in Mathcad have alternative display symbols (e.g., ·, •, and × for multiplication:); the assignment operator, :=, will be used as an example.

To bring up the Worksheet Options dialog, use Tools/Worksheet Options ... from the Mathcad menu:

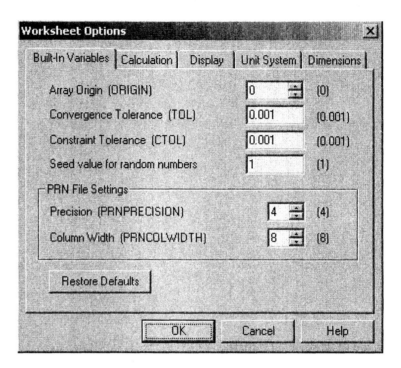

Click on the Display tab to change the way operators are displayed:

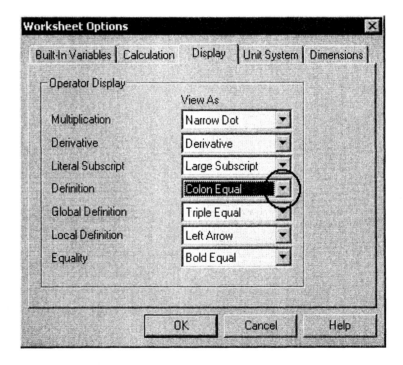

To change the way assignment (i.e., variable definition) is indicated, click on the drop-down list on the Definition line, and select Equal:

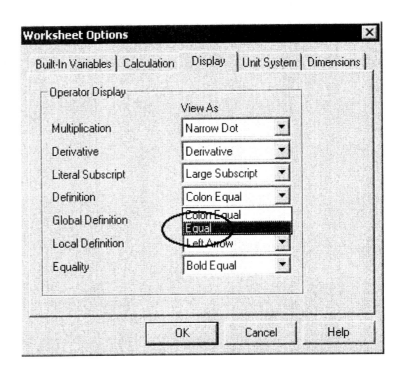

Now, every := on your worksheet will be displayed without the colon before the equal sign. The following lists show a few examples of what is displayed before and after changing the assignment operator:

Before changing the assignment operator   *After changing the assignment operator*

```
R := 3•cm                                        R = 3•cm
L := 12•cm                                       L = 12•cm
A := 2•π•R²+2•R•π•L                              A = 2•π•R²+2•R•π•L
A = 282.7cm²                                     A = 282.7cm²
```

Note: Changing the assignment operator changes the way equations are displayed, but not the way they are entered; the equations for R and L were entered by using the colon [:] key.

2.5 WORKING WITH UNITS

Mathcad supports units. Its ability to automatically handle *unit conversions* is a *very* nice feature for engineering calculations, since the number of required unit conversions can be considerable. Mathcad handles units by storing all values in a base set of units (SI by default, but you can change it):

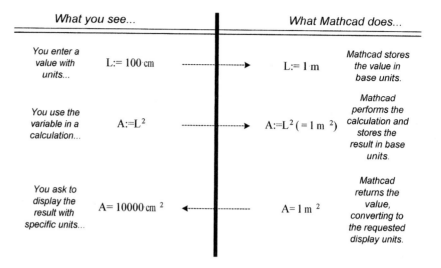

Values are converted from the units you enter to the base set before the value is stored. Then, when a value is displayed, the conversions required to give the requested units are automatically performed. Mathcad's unit handling works very well for most situations, but it does have some limitations, which will be discussed shortly.

Mathcad supports the following *systems of units*:

- SI—default units (meter, kilogram, second, etc.);
- MKS—meter, kilogram, second;
- CGS—centimeter, gram, second;
- US—foot, pound, second;
- none—disables all built-in units, but user-defined units still work.

Within each system, certain *dimensions* are supported. But not all dimensions are supported in all systems. The following table shows the dimensions the various systems support in Mathcad:

DIMENSION	SYSTEM				
	SI	MKS	CGS	US	None
Mass	✓	✓	✓	✓	✓
Length	✓	✓	✓	✓	✓
Time	✓	✓	✓	✓	✓
Current	✓	no	no	no	no
Charge	no	✓	✓	✓	✓
Temperature	✓	✓	✓	✓	✓
Luminosity	✓	no	no	no	no
Substance	✓	no	no	no	no

If you plan to work with current (typical units, amps), luminosity (typical units, candelas) or substance (typical units, moles), then you want to use the default SI units. If these units are not important to you, you have more system options available. You set the system of units to be used as base units from the menu bar: Tools/Worksheet Options.../Unit System.

The following are some common *unit abbreviations*, by category:

mass	kg, gm, lb
length	m, cm, ft
time	sec, hr
current	amp
charge	coul
temperature	K, R
substance	mole
volume	liter, gal, galUK (Imperial gallon)
force	N, dyne, lbf
pressure	Pa, atm, torr, in_Hg, psi
energy and work	joule, erg, cal, BTU
power	watt, kW, hp

CAUTIONS ON UNIT NAMES

g—predefined as gravitational acceleration ($9.8 \, m/s^2$), not grams;
gm—unit name for grams;
R—unit name for °R (degrees Rankine), not the gas constant;
mole—gram mole, defined in the SI system only.

You can overwrite the symbols used as predefined unit names by declaring them as variables in a worksheet. For example, if you included the statement

```
m := 1•kg
```

in a worksheet, the symbol 'm' would become a variable and no longer represent meters. Redefining unit names is usually not a good idea.

These units are only a small subset of the predefined units in Mathcad. A list of available units may be obtained by pressing [Ctrl-u]. You can select units from the list or simply type the unit name (using Mathcad's abbreviation).

For example, you could find the area of a circle with a radius of 7 cm by using the following equations:

```
r := 7•cm
Area := π•r²
```

In the first line, the 7 is multiplied by the unit name, 'cm'; then the area is computed on the second line. The result is displayed as

```
Area| = 0.015•m² ▮
```

The box around this equation and the vertical bar at the right side of Area indicate that we haven't completed editing the equation; that is, we haven't pressed [Enter] yet. As soon as you press [=], Mathcad shows the value currently assigned to the variable Area in the base units (SI, by default). A *placeholder* is also shown to the right of the units (only while the equation is being edited), so that you can request that the value be displayed in different units. If you want the result displayed in cm², click on the placeholder and then enter [c] [m] [^] [2] [Enter]. The result is displayed in the requested units:

```
Area = 153.938•cm²
```

You can put any defined (predefined, or defined part of the worksheet) unit in the placeholder. If the units you enter have the wrong dimensions, Mathcad will make that apparent by showing you the "leftover" dimensions in base units. For example, if you had placed 'atm' (atmospheres) in the placeholder, the dimensions would be quite wrong, and the displayed result would make this apparent:

```
Area = 1.519•10⁻⁷•kg⁻¹•m³•s² ∘atm
```

2.5.1 Defining a New Unit

New units are defined in terms of predefined units. For example, a commonly used unit of viscosity is the centipoise, or cP. The poise is predefined in Mathcad, but the centipoise is not. The new unit can be defined like this:

$$cP := \frac{poise}{100}$$

Once the new unit has been defined, it can be used throughout the rest of the worksheet. Here, the viscosity of honey is defined as 10,000 cP, and then the value is displayed by using another common unit for viscosity, Pascal · seconds:

```
visc_honey := 10000•cP
visc_honey = 10∘Pa•s
```

PRACTICE!

Try these obvious unit conversions:

a. 100 cm → m
b. 2.54 cm → in
c. 454 gm → lb

Now try these less obvious conversions:

a. 1 hp → kW
b. 1 liter · atm → joule
c. 1 joule → watt (This is an invalid conversion. How does Mathcad respond?)

PROFESSIONAL SUCCESS

First, make sure your method [or design/computer code/Mathcad worksheet] works correctly on a test case with a known answer. Then try your method [design/computer code/Mathcad worksheet] on a new problem.

In the last "Practice"! box, you were asked to try Mathcad on some obvious cases before trying some less obvious ones. Testing against a known result is a standard procedure in engineering. If you get the right answer in the test case, you have increased confidence (though not total confidence) that your method is working. If you get the wrong answer, it is a lot easier to find out what went wrong using the test problem than it is to try to fix the method and solve the real problem simultaneously.

You'll see this technique used throughout the text.

2.5.2 Editing the Units on a Value or Result

Once you have entered units, simply edit the unit name to change them. Click on the unit name, delete characters as needed, and then type in the desired units.

2.5.3 Limitations to Mathcad's Units Capabilities

- Mathcad's unit conversions must be multiplicative—no additive constants can be used in unit conversions. This means that Mathcad can convert kelvins to °R (degrees Rankine), but cannot convert °C to kelvins (add 273) or °C to °F (add 32). Similarly, Mathcad can handle absolute pressure conversions, but cannot automatically convert gauge pressures to absolute (add the barometric pressure).
- Some of Mathcad's built-in functions do not support, or do not fully support, units. For example, the linear regression function `linfit( )` does not accept values with units. The iterative solver (`given-find` solve block) does allow units, but if you are solving for two or more variables simultaneously, all of the variables must have the same units.
- Mathcad's built-in graphics always display the values in the base (stored) units.
- Only the SI system of units fully supports moles. In the other systems, Mathcad allows you to use the term "mole" as a unit, but does not consider it a dimension and does not display it when presenting units. This means that *you should use the SI system if you plan to work with moles*.
- The mole defined in Mathcad's SI system is the gram-mole. I like to make this obvious by defining a new unit, the gmol:

    ```
    gmol := mole
    ```

    You can define the other commonly used molar units as well:

    ```
    kmol  := 1000•mole
    lbmol := 453.593•mole
    ```

    With these definitions, Mathcad can convert between the various types of moles (but only if you are using the SI system of units.)

APPLICATIONS: DETERMINING THE CURRENT IN A CIRCUIT

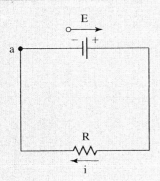

Consider a simple circuit containing a 9-volt battery and a 90-ohm resistor, as shown in the accompanying diagram. What current would flow in the circuit? To solve this problem, you need to know Ohm's law,

$$V = iR,$$

and Kirchhoff's law of voltage,

For a closed (loop) circuit, the algebraic sum of all changes in voltage must be zero.

Kirchhoff's law is perhaps easier to understand if you consider the following analogy:

> If you are hiking in the hills and you end up back at the same spot you started from (a loop trail), then the sum of the changes in elevation must be zero: You must end up at the same elevation at which you started.

The electrons moving through the circuit from point 'a' may have their electric potential (volts) increased (by the battery) or decreased (by the resistor), but if they end up back where they started from (in a loop circuit), they must end up with the same potential they started with. We can use Kirchhoff's law to write an equation describing the voltage changes through the circuit.

Starting at an arbitrary point 'a' and moving through the circuit in the direction of the current flow the voltage is first raised by the battery ($V_B = E$) and then lowered by the resistor (V_R). These are the only two elements in the circuit, so

$$V_B - V_R = 0$$

by Kirchhoff's law. Now, we know that the battery raises the electric potential by 9 volts, and Ohm's law relates the current flowing through the resistor (and the rest of the circuit) to the resistance, 90 ohms. Thus,

$$V_B - iR = 0.$$

The Mathcad equations needed to calculate the current through the circuit look like this:

$$V_B := 9 \cdot \text{volt}$$
$$R := 90 \cdot \text{ohm}$$
$$i := \frac{V_B}{R}$$
$$i = 0.1 \circ \text{amp}$$

2.6 CONTROLLING HOW RESULTS ARE DISPLAYED

The Result Format dialog allows you to control the way individual numbers, matrices, and units are displayed. If you have not selected a result when you change a setting in the Result Format dialog, you will change the way all results on the worksheet will be displayed. However, if you select a value before you open the Result Format dialog, the changes you make in the Result Format dialog will affect only that one result.

The Result Format dialog can be opened in two ways:

1. from the menu, using Format / Result ..., or
2. by double-clicking on a displayed result.

2.6.1 Controlling the Way Numbers Are Displayed

Mathcad tries to present results in a readable form, with only a few decimal places displayed. You may want to change the way results are displayed, in any of following ways:

1. Format: You can use General, Decimal, Scientific, Engineering, or Fraction format. (The default is General format, which displays small values without exponents, but switches to scientific notation for larger values.)
2. Number of decimal places displayed (three by default).
3. Whether or not to display trailing zeros. (By default, they are not displayed.)
4. Exponential threshold (for General format only): You can indicate how large values can get before Mathcad switches to scientific notation.

You can bring up the Result Format dialog by double-clicking on a displayed result. As an example, we'll use $B = 1/700$, a very small value with lots of decimal places:

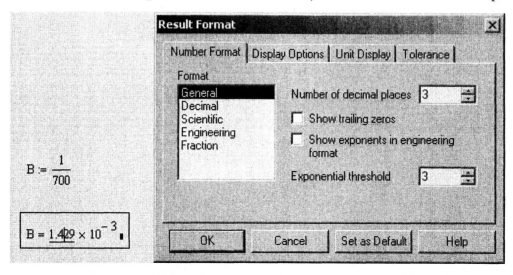

The Mathcad defaults are shown in the Result Format dialog:
- general format;
- three decimal places;
- trailing zeros not shown;
- exponential threshold of 3 (i.e., 999 displayed as decimal, 1,000 displayed in scientific notation)

If you were to select Decimal format, the result would be displayed in standard notation, but still with only three decimal places:

$$B := \frac{1}{700}$$
$$B = 0.001$$

To see more decimal places, change the number of displayed decimal places from 3 to some larger value, say, 7:

$$B := \frac{1}{700}$$
$$B = 0.0014286$$

The following table compares general, scientific, and engineering formats:

DEFINITION X	GENERAL FORMAT	SCIENTIFIC FORMAT	ENGINEERING FORMAT
$x := \begin{pmatrix} 10 \\ 100 \\ 1000 \\ 10000 \\ 100000 \\ 1000000 \end{pmatrix}$	$x = \begin{pmatrix} 10.000 \\ 100.000 \\ 1.000 \times 10^3 \\ 1.000 \times 10^4 \\ 1.000 \times 10^5 \\ 1.000 \times 10^6 \end{pmatrix}$	$x = \begin{pmatrix} 1.000 \times 10^1 \\ 1.000 \times 10^2 \\ 1.000 \times 10^3 \\ 1.000 \times 10^4 \\ 1.000 \times 10^5 \\ 1.000 \times 10^6 \end{pmatrix}$	$x = \begin{pmatrix} 10.000 \times 10^0 \\ 100.000 \times 10^0 \\ 1.000 \times 10^3 \\ 10.000 \times 10^3 \\ 100.000 \times 10^3 \\ 1.000 \times 10^6 \end{pmatrix}$

General format displays small values without a power, but switches large values (greater than the exponential threshold) to scientific notation. Scientific format uses scientific notation for values of any size. Engineering format is similar to scientific notation, except that the powers are always multiples of 3 ($-6, -3, 0, 3, 6, 9$, etc.).

2.6.2 Controlling the Way Matrices Are Displayed

By default, Mathcad shows small matrices (arrays) in their entirety, but switches to scrolling tables when the matrices get large. You can change the way a matrix result is displayed by double-clicking on the matrix:

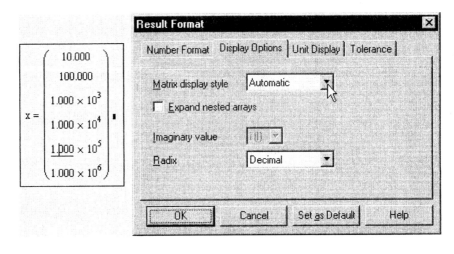

The x matrix used in this example is a pretty small matrix, so the Automatic selection (default) displays the result as a matrix. You can change the display style using the drop-down list and selecting Table:

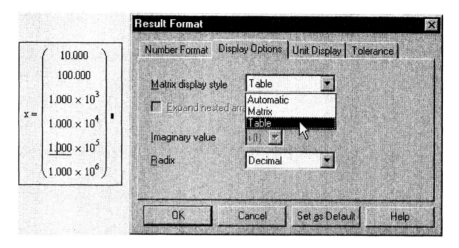

When you click on the OK button, the x matrix is displayed as a scrolling table:

$$x = \begin{array}{|c|c|} \hline & 0 \\ \hline 0 & 10.000 \\ \hline 1 & 100.000 \\ \hline 2 & 1{,}000 \cdot 10^3 \\ \hline 3 & 1{,}000 \cdot 10^4 \\ \hline 4 & 1{,}000 \cdot 10^5 \\ \hline 5 & 1{,}000 \cdot 10^6 \\ \hline \end{array}$$

2.6.3 Controlling the Way Units Are Displayed

By default, Mathcad tries to simplify units whenever possible. Because of this, the units on the result in the following example are newtons rather than $kg \cdot m / s^2$:

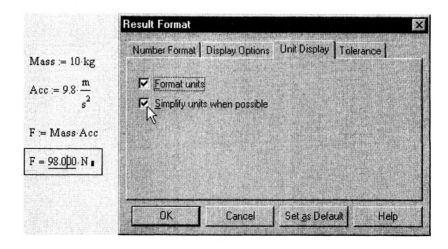

If you do not want Mathcad to simplify the units for a particular result, double-click on that result to bring up the Result Format dialog. Then, clear the "Simplify units when possible" check box. When you leave the dialog box, the result will be displayed without simplifying the units:

$$\text{Mass} := 10 \cdot \text{kg}$$
$$\text{ACC} := 9.8 \cdot \frac{m}{s^2}$$
$$F := \text{Mass} \cdot \text{Acc}$$
$$F = 98.000 \cdot \frac{\text{kg m}}{s^2}$$

The "Format units" check box indicates that units should be displayed in "common" form. If the "Format units" check box is cleared, the result would be displayed as

$$F = 98.000 \cdot \text{kg m s}^{-2}$$

2.7 ENTERING AND EDITING TEXT

Mathcad defaults to equation edit mode, so if you just start typing, Mathcad will try to interpret your entry as an equation. If you type a series of letters and then a space, Mathcad will recognize that you are entering text and will switch to *text edit mode* and create a *text region*. Or you can tell Mathcad you want to enter text by pressing the double-quote key ["].

To create a text region, position the edit cursor (crosshair) in the blank portion of the worksheet where you want the text to be placed, and then press ["] (the double-quote key). A small rectangle with a vertical line (the insert bar) inside appears on the worksheet, indicating that you are creating a text region. The text region will automatically expand as you type, until you reach the page margin (the vertical line at the right side of the worksheet. Then the text will automatically wrap to the next line. The result looks as follows:

> This text region will expand until reaches the page margin, then the text will automatically wrap...

You can change the size or location of an existing text region by clicking on the text to select the region and display the border. Then, drag the border to move the box, or drag one of the handles to change the size of the box. If you change the width of a text region, the text will wrap as necessary to fit in the new width.

Once you have a block of text in a text region, you can select all or part of the text with the mouse and then use the formatting buttons on the Format Toolbar. These buttons allow you to make the font boldface, add italics, or underline the text. For example, to create a heading for some global-warming calculations, you might type in some text and then select "Global Warming":

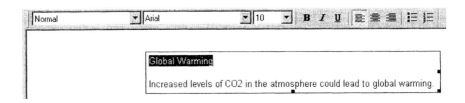

Then increase the font size of the selected text by using the Format Toolbar:

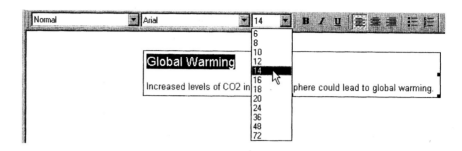

Finally, make the font in the heading boldface by using the Format Toolbar:

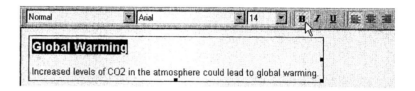

You can also create *sub-* or *superscripts* within the text region by selecting the characters to be raised or lowered; then press the Subscript or Superscript buttons on the Format Toolbar:[2]

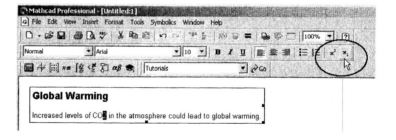

After these changes, the worksheet heading looks like this:

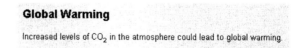

Mathcad gives you a great deal of control over the appearance of your text, and adding headings and notes to your worksheets can make them much easier for others to read and understand.

---

[2]The subscript and superscript buttons are new in Mathcad 11. In previous versions, select the text to be raised or lowered, then use Format/Text..., and select either the Superscript or Subscript buttons on the Text Format dialog.

PROFESSIONAL SUCCESS

Document your results.
Taking the time to document your worksheets by adding headings and comments can have a big payoff later (sometimes years later), when you need to perform a similar calculation and want to refer to your worksheets to see how it was done.

The skills you develop and the knowledge you acquire in your academic and professional careers combine to form the resource base, or expertise, that you bring to a new job or project. But when your resource base gets stale, you can spend a lot of time relearning things you once knew. Plan ahead and make the relearning as easy as possible by making your worksheets easy to understand.

Note: If you want to make a change to all text regions—italicizing all the text so that it is easier to see the difference between your text regions and equations, for example—then change the *style* used to display all text regions. To modify the text style sheet, select Format/Style .../Normal/Modify/Font from the menu bar. This will bring up a dialog box describing the Normal style (used for regular text). If you click on the italic property and close the dialog box, all text created using the Normal style will be in italics. (Format changes to individual text boxes or selected characters override the style sheet. If you select one word in a text box and remove the italics using the Format Bar, that individual change would override the style sheet, and the italics would be removed from that word.)

2.7.1 Sizing and Moving a Text Region or an Equation Region

If you click outside of a region and drag the mouse into the region, Mathcad will show you the size and location of the region by drawing a line around it. For resizable regions, like text regions, Mathcad will also show three small black squares on the edges of the region, called *handles*. Grab a handle with the mouse, and drag it to change the size or shape of a region. (Some regions automatically resize, so they might snap back again when you release the handle.) To move a single selected region, move the mouse over the border until it changes to a hand symbol. While the hand symbol is displayed, drag the border (and the region) to the new location on the worksheet. Or, hold the shift key down and click anywhere on a region. (The selected region will be displayed with a dashed line.) You can then move the region with the mouse or the cursor keys on the keyboard. If multiple regions are selected, dashed lines are drawn around each region. They cannot be resized, but they can be moved together by clicking inside any of the selected regions and dragging all the selected regions to a new location.

2.7.2 Inserting Equations inside Text Regions

You can move an existing equation into a text region so that the equation is read as part of the text. When equations are moved inside text regions, they are still functioning equations in the Mathcad worksheet and are evaluated just like any other equation.

As an example, consider a force calculation with a paragraph describing Newton's second law:

```
M := 10·kg
a := 9.8·m/s²
F := M·a
F = 98 N
```

The relationship between force, mass, and acceleration was defined by Isaac Newton. The commonly used version of the equation is called Newton's second law.

The equation for calculating force can be moved inside the paragraph, and it will still calculate the force:

$$M := 10 \cdot kg$$
$$a := 9.8 \cdot \frac{m}{s^2}$$
$$F = 98 \ N$$

The relationship between force, mass, and acceleration was defined by Isaac Newton. The commonly used version of the equation, $F := M \cdot a$, is called Newton's second law.

2.8 A SIMPLE EDITING SESSION

We will use one of the examples mentioned earlier, namely, determining the surface area of a cylinder, to demonstrate how to solve a simple problem using Mathcad. Even in this simple example, the following Mathcad features will be used:

- text editing and formatting;
- variable definitions (r and L);
- equation definition;
- exponentiation (r^2);
- displaying the result (=);
- working with units.

2.8.1 Statement of the Problem

Determine the surface area of a cylinder with a radius of 7 cm and a length of 21 cm.

The solution will be presented with a step-by-step commentary. For this simple example, the complete Mathcad worksheet will be shown as it develops at each step.

Step 1: Use a Text Region to Describe the Problem

1. Position the edit cursor (crosshair) near the top of the blank worksheet and click the left mouse button.
2. Press ["] to create a text region.
3. Enter the statement of the problem.
4. Click outside of the text region when you are done entering text.

The result might look something like this (italic text is not the Mathcad default, but adding italics can help differentiate text regions from equations):

Statement of the Problem
    Determine the surface area of a cylinder with a radius of 7 cm and a length of 21 cm.

Step 2: Enter The Known Values of Radius and Length, With Appropriate Units

1. Position the edit cursor below the text region and near the left side of the worksheet. Click the left mouse button.
2. Define the value of r by typing [r] [:] [7] [*] [c] [m] [Enter].
3. Similarly, position the edit cursor, and enter the value of L (21 cm).

The positions of the r and L equations are arbitrary, but they might look like the following:

Statement of the Problem
    Determine the surface area of a cylinder with a radius of 7 cm and a length of 21 cm.

$$r := 7 \cdot cm$$
$$L := 21 \cdot cm$$

Step 3: Enter The Equation for Computing The Surface Area of a Cylinder

1. Position the edit cursor below or to the right of the definitions of r and L.
2. Enter the area variable name, A_{cyl}, by using a text subscript, as [A] [.] [c] [y] [l].
3. Press [:] to indicate that you are defining the A_{cyl} variable.
4. Enter the first term on the right side $(2\pi r^2)$ as [2] [*] [Ctrl-Shift-p] [*] [r] [^] [2].[3]
5. Press [Space] to select the r^2 in the $2\pi r^2$ term.
6. Press [+] to add to the selected term.
7. Enter the second term on the right side $(2\pi rL)$ as [2] [*][Ctrl-Shift-p] [*] [r] [*] [L].
8. Press [Enter] to conclude the equation entry.

The result will look like this:

> **Statement of the Problem**
> Determine the surface area of a cylinder with a radius of 7 cm and a length of 21 cm.
>
> r := 7•cm
> L := 21•cm
> A_{cyl} := 2•π•r²+2•π•r•L

Step 4: Display The Result in The Desired Units

1. Position the edit cursor below or to the right of the A_{cyl} equation.
2. Ask Mathcad to display the result of the calculation by typing [A] [.] [c] [y] [l] [=].
3. Click on the units placeholder (to the right of the displayed base units).
4. Enter the desired units (e.g., cm^2).
5. Press [Enter] to conclude the entry.

The worksheet will look like this:

> **Statement of the Problem**
> Determine the surface area of a cylinder with a radius of 7 cm and a length of 21 cm.
>
> r := 7•cm
> L := 21•cm
> A_{cyl} := 2•π•r²+2•π•r•L
> A_{cyl} = 1.232•10³•cm²

PRACTICE!

Calculate the surface area and volume of
  a. a cube 1 cm on a side.
  b. a sphere with a radius of 7 cm.

Equations for the sphere are $A = 4\pi r^2$ and $V = (4/3)\pi r^3$.

---

[3] The keyboard shortcut is [Ctrl-Shift-p] in recent versions of Mathcad and [Ctrl-p] in version 7.

APPLICATION: NANOTECHNOLOGY FOR COMPUTER CHIPS

The number of transistors on computer chips has been growing exponentially for the past 30 years, doubling approximately every 18 months. This has fueled the rapid increase in computer power and simultaneous decrease in costs that we have witnessed over the past decade. But those transistors are formed on chips using photolithographic methods; essentially they are printed on silicon using light. The wavelength of the light source imposes a fundamental limitation on the minimum size of the transistors, and chip manufacturers have been switching to higher frequency, shorter wavelength light sources to get the printed line width down to 0.14 μm (or 140 nm). But the chip manufacturers know that they are rapidly running out of opportunities to obtain ever smaller line widths using photolithography. The search is on for radically different approaches to chip building.

One of the radical new approaches that shows promise is the use of semiconducting carbon *nanotubes* as transistors. These nanotubes are cylinders made of carbon atoms, approximately 5 to 10 atoms in diameter. Nanotubes are currently being produced and are being studied for a wide range of applications, including potential use as transistors.

As a quick (and very superficial) estimate of the potential for nanotubes to increase the power of computer chips, let's calculate how many nanotubes could fit on a 220 mm² computer chip.

First, we need to define millimeters, micrometers, and nanometers in Mathcad:

$$mm := 10^{-3} \cdot m \qquad \mu m := 10^{-6} \cdot m$$
$$nm := 10^{-9} \cdot m$$

Nanotubes are approximately 1 nm in diameter and perhaps 200 to 500 nm in length. If the nanotube is attached to the surface of a chip, it would occupy an area of up to 500 nm². Thus, we have

$$D := 1 \cdot nm \qquad L := 500 \cdot nm$$
$$A_{nt} := D \cdot L \qquad A_{nt} = 500 nm^2$$

A recent, commonly produced computer chip held about 50 million transistors on a silicon chip with an area of 220 mm². If you pack the surface with nanotubes, you could fit 440 billion nanotubes in the same area, a result derived as follows:

$$A_{chip} := 220 \, mm^2 \qquad A_{chip} = 2.2 \times 10^{14} \, nm^2$$
$$N_{tubes} := \frac{A_{chip}}{A_{nt}} \qquad N_{tubes} = 440 \times 10^9$$

While you probably would not be able to pack the nanotubes so tightly, if you used just a tenth of the tubes in the same area, you would still increase the number of transistors per chip by a factor of 1000 compared with today's chips.

SUMMARY

In this chapter, we learned the basics of working with Mathcad: how the screen is laid out, how to enter equations and text, and how Mathcad handles units.

MATHCAD SUMMARY

Four Kinds of Equal Signs

:=	Assigns a value or the result of a calculation to a variable.
=	Displays a value or the result of a calculation.
=	Symbolic equality; shows the relationship between variables.
≡	Global assignment; these are evaluated before the rest of the worksheet.

Predefined Variables

π	3.141592...	Press [Ctrl-Shift-p] or choose π from the Greek Symbols Toolbar.
e	2.718281...	
g	9.8 m/s²	
%		multiplies the displayed value by 100 and displays the percent symbol.

Entering Equations

+, −	addition and subtraction.	
*, /	multiplication and division.	Press [Shift-8] for the multiplication symbol.
^	exponentiation.	Press [Shift-6] for the caret symbol.
[Space]	enlarges the currently selected region—used after typing in exponents and denominators.	
[Insert]	moves the vertical edit cursor between the front and back of a selected region—used when you need to delete an operator to the left of a selected region.	
[.]	text subscript.	
[[ ]	index subscript—used to indicate a particular element of an array.	

Operator Precedence

^	Exponentiation is performed before multiplication and division.
*, /	Multiplication and division are performed before addition and subtraction.
+, −	Addition and subtraction are performed last.

Text Regions

["]	Creates a text region—if you type in characters that include a space (i.e., two words), Mathcad will automatically create a text region around these characters.

KEY TERMS

anchor point	insert bar	style
assignment operator	matrix	symbolic equality
dimension	operator	text region
edit cursor	operator precedence	text subscript
equation region	placeholder	toolbar
global assignment	predefined variable	unit system
index (array) subscript	radix	worksheet

Problems

1. **UNIT CONVERSIONS**

   a. speed of light in a vacuum — 2.998×10^8 m/sec to miles per hour.

   b. density of water at room temperature — 62.3 lb/ft^3 to kg/m^3.

   c. density of water at 4°C — 1,000 kg/m^3 to lb/gal.

   d. viscosity of water at room temperature (approx.) — 0.01 poise to lb/ft sec. 0.01 poise to kg/m sec.

   e. ideal gas constant — 0.08206 L·atm/mole·K to joules/mole·K.

   Note: The "lb" in parts b, c, and d is a pound mass, and "mole" in part e is a gram mole.

2. **VOLUME AND SURFACE AREA OF A SPHERE**

   Calculate the volume and surface area of a sphere with a radius of 3 cm. Use

   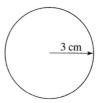

   $$V = \frac{4}{3}\pi r^3$$

   and

   $$A = 4\pi r^2.$$

3. **VOLUME AND SURFACE AREA OF A TORUS**

   A doughnut is shaped roughly like a torus. The surface area and volume functions for a torus are

   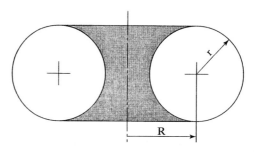

   $$A = 4\pi^2 R\, r$$

   and

   $$V = 2\pi^2 R\, r^2,$$

   respectively. Calculate the surface area and volume of a doughnut with $R = 3$ cm and $r = 1.5$ cm.

4. **IDEAL GAS BEHAVIOR, I**

   a. A glass cylinder fitted with a movable piston contains 5 grams of chlorine gas. When the gas is at room temperature (25°C), the piston is 2 cm from the bottom of the container. The pressure on the gas is 1 atm. What is the volume of gas in the glass cylinder (in liters)?

   b. The gas is heated at constant pressure until the piston is 5 cm from the bottom of the container. What is the final temperature of the chlorine (in kelvins)? Assume chlorine behaves as an ideal gas under these conditions. The molecular weight of chlorine is 35.45 grams per mole.

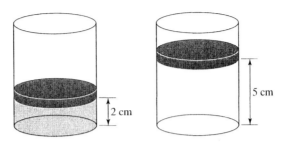

The ideal gas equation is written as

$$PV = NRT$$

Where P, V, and T represent the pressure, volume, and temperature of N moles of gas. R is the ideal gas constant. The value of R depends on the units used on the various variables. A commonly used value for R is 0.08206 liter · atm/mole · K.

5. **IDEAL GAS BEHAVIOR, II**

   a. Chlorine gas is added to a glass cylinder 2.5 cm in radius, fitted with a movable piston. Gas is added until the piston is lifted 5 cm off the bottom of the glass. The pressure on the gas is 1 atm, and the gas and cylinder are at room temperature (25°C). How many moles of chlorine were added to the cylinder?

   b. Pressure is applied to the piston, compressing the gas until the piston is 2 cm from the bottom of the cylinder. When the temperature returns to 25°C, what is the pressure in the cylinder (in atm)?

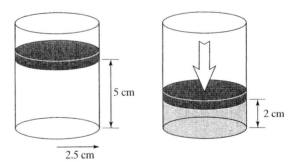

Assume that chlorine behaves as an ideal gas under these conditions. The molecular weight of chlorine is 35.45 grams per mole.

6. **RELATING FORCE AND MASS**

Fundamental to the design of the bridge is the relation between the mass of the deck and the force on the wires supporting the deck. The relationship between force and mass is called *Newton's law* and in physics courses is usually written as

$$F = ma.$$

In engineering courses, we often build in the gravitational constant, g_c, to help keep the units straight:

$$F = m\frac{a}{g_c}.$$

Note that g_c has a value of 1 (no units) in SI and a value of 32.174 ft·lb/lb$_f$·s^2 in American Engineering units. If the mass is being acted on by gravity, Newton's law can be written as

$$F = m\frac{g}{g_c},$$

where $g = 9.8$ m/s^2 or 32.174 ft/s^2 is the acceleration due to gravity.

Note: While the acceleration due to gravity, g, is predefined in Mathcad, g_c is not. If you want to use g_c, it can easily be defined in your worksheet with the use of the SI value, then let Mathcad take care of the unit conversions. The definition is simply

$$g_c := 1$$

a. If a 150-kg mass is hung from a hook by a fine wire (of negligible mass), what force (N) is exerted on the hook? See accompanying diagram.

b. If the mass in part (a) were suspended by two wires, the force on the hook would be unchanged, but the tension in each wire would be halved. If the duty rating on the wire states that the tension in any wire should not exceed 300 N, how many wires should be used to support the mass?

7. SPRING CONSTANTS

Springs are so common that we hardly even notice them, but if you are designing a component that needs a spring, you have to know enough about them to specify them correctly. Common springs obey *Hooke's law* (if they are not overstretched), which simply states that the spring extension, x (the amount of stretch), is linearly related to the force exerted on the spring, F; that is,

$$F = kx,$$

where k is the linear proportionality constant, called the *spring constant*. Use Mathcad's unit capabilities to determine the spring constants in N/m for the following springs:

  a. extended length, 12 cm; applied force, 800 N.
  b. extended length, 0.3 m; applied force, 1200 N.
  c. extended length, 1.2 cm; applied force, 100 dynes.
  d. extended length, 4 inches; applied force, 2000 lb_f.

Note: The preceding equation is actually only a simplified version of Hooke's law that is applicable when the force is in the direction of motion of the spring. Sometimes you will see a minus sign in Hooke's law. It all depends on whether the force being referred to is being applied to the spring or is within the spring, restraining the applied force.

8. SPECIFYING A SPRING CONSTANT

The backrest of a chair is to be spring loaded to allow the chair to recline slightly. The design specifications call for a deflection of no more than 2 inches when a 150-lb person leans 40% of his or her body weight on the backrest. (Assume that there are no lever arms between the backrest and the spring to account for in this problem.)

  a. Use Newton's law (described in Problem 6) to determine the force applied to the backrest when a 150-lb person leans 40% of his or her body weight against it.
  b. Determine the constant required for the spring.
  c. If a 200-lb person puts 70% of his or her body weight on the backrest, what spring extension would be expected? (Assume that the applied force is still within the allowable limits for the spring.)

9. SIMPLE HARMONIC OSCILLATOR

If a 50-gram mass is suspended on a spring and the spring is stretched slightly and released, the system will oscillate. The period, T, and natural frequency, f_n, of this simple harmonic oscillator can be determined using the formulas

$$T = \frac{2\pi}{\sqrt{k/m}}$$

and

$$f_n = \frac{1}{T},$$

where k is the spring constant and m is the suspended mass. (This equation assumes that the spring has negligible mass.)

a. If the spring constant is 100 N/m, determine the period of oscillation of the spring.

b. What spring constant should be specified to obtain a period of 1 second?

10.   DETERMINING THE CURRENT IN A CIRCUIT

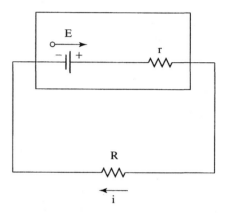

A real battery has an internal resistance, often shown by combining a cell symbol and a resistor symbol, as illustrated in the preceding figure. If a 9-volt battery has an internal resistance, r, of 15 ohms and is in a circuit with a 90-ohm resistor, what current would flow in the circuit?

Note: When resistors are connected in series, the combined resistance is simply the sum of the individual resistance values.

11.   RESISTORS IN PARALLEL

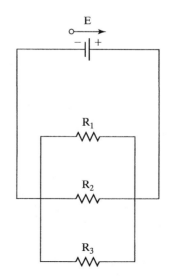

When multiple resistors are connected in parallel (see preceding diagram), the equivalent resistance of the collection of resistors can be computed as

$$\frac{1}{R_{eq}} = \sum_{i=1}^{N} \frac{1}{R_i},$$

where N is the number of resistors connected in parallel:

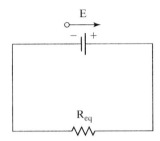

Compute the equivalent resistance, R_{eq}, in the circuit, and determine the current that would flow through the equivalent resistor. The data are as follows:

$$E = 9 \text{ volts};$$
$$R_1 = 20 \text{ ohms};$$
$$R_2 = 30 \text{ ohms};$$
$$R_3 = 40 \text{ ohms}.$$

12. **BUILDING NANOTUBES**

    Carbon has an atomic radius of 77.2 pm (pm means picometers, 1 pm = 1×10^{-12} m). Approximately how many carbon atoms are required to build a nanotube 1 nm in diameter by 300 nm long?

13. **AVERAGE FLUID VELOCITY**

    When a fluid is flowing in a pipe, the rate and direction of flow of the fluid — the fluid velocity—is not the same everywhere in the pipe. The fluid near the wall moves more slowly than the fluid near the center of the pipe, and things that cause the flow field to bend, such as obstructions and bends in the piping, cause some parts of the flow to move faster and in different directions than other parts of the flow. For many calculations, the details of the flow pattern are not important, and an *average fluid velocity* can be used in design calculations. The average velocity, V_{avg}, can be determined by measuring the volumetric flow rate, Q, and dividing that by the cross-sectional area of the pipe, A_{flow}:

    $$V_{avg} = \frac{Q}{A_{flow}}.$$

    a. If 1,000 gallons per minute (gpm) of water are flowing through a 4-inch (inside diameter) pipe, what is the average velocity of the water in the pipe?

    b. If the flow passes into a section of 2-inch pipe, what is the average velocity in the smaller pipe?

c. A common rule of thumb for trying to keep pumping costs low is to design for an average fluid velocity of about 3 ft/sec. What pipe diameter is required to obtain this average velocity with a volumetric flow rate of 1,000 gpm?

Note: You may need to take a square root in part c. You can use either an exponent of 0.5 or Mathcad's square-root operator, which is available by pressing [\] (backslash).

14. ORIFICE METER

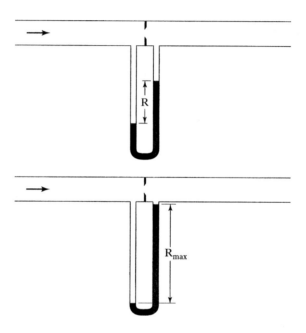

An *orifice meter* is a commonly used flow measuring device. The flow must squeeze through a small opening (the orifice), generating a pressure drop across the orifice plate. The higher the flow rate, the higher is the pressure drop, so if the pressure drop is measured, it can be used to calculate the flow rate through the orifice.

In the past, mercury manometers were often used to measure the pressure drop across the orifice plate. The manometer reading, R, can be related to the pressure drop, ΔP, by the differential manometer equation

$$\Delta P = (\rho_m - \rho)\frac{g}{g_c}R,$$

where ρ_m is the density of the manometer fluid, ρ is the density of the process fluid in the pipe, g is the acceleration due to gravity, g_c is the gravitational constant (used to help manage units in the American Engineering system), and R is the difference in height of the manometer fluid on the two sides of the manometer, which is called the manometer reading.

a. Given the following data, what is the pressure drop (atm) across the manometer when the manometer reading is $R = 32$ cm?

$\rho_m = 13{,}600 \text{ kg/m}^3$;
$\rho = 1{,}000 \text{ kg/m}^3$;
$g = 9.80 \text{ m/sec}^2$;
$g_c = 1 \text{ (SI)}$;

b. If the flow rate gets too high, the mercury in the manometer can be washed out of the device into the pipe downstream of the orifice plate. To avoid this, the orifice must be designed to produce a pressure drop smaller than the maximum permissible ΔP for the manometer. If $R_{max} = 70$ cm, what is the maximum permissible ΔP (atm) for the manometer?

15. DESIGNING AN IRRIGATION SYSTEM

In arid areas, one cannot rely on periodic rains to water crops, so many farmers depend on irrigation systems. Center-pivot irrigation systems are a type of system that is commonly used. These systems have a 1/4-mile-long water pipe, on wheels, that rotates (pivots) around the water source. The result is an irrigated circle with a diameter of 1/2 mile, centered on the source. From an airplane, you can easily see where center-pivot systems are in use.

In the past, one large pulsating sprinkler was used for each wheeled section (120 feet) of pipe, and the sprinkler head was placed high to get as wide a coverage as possible. In an attempt to utilize the water better, these systems have been radically reengineered. The large pulse sprinkler heads have been replaced by a series of small water spray nozzles 20 feet apart, and these have been placed much lower, just above crop level, to reduce the impact of wind on the water distribution pattern.

a. How many acres of cropland would be irrigated using the center-pivot system just described?

b. An acre-inch of water is the volume of water needed to cover an acre of ground with water to a depth of one inch (assuming no ground infiltration). How many acre-inches of water are required to provide one inch (depth) of water to the field? How many gallons is that?

c. If the system makes a complete rotation every 40 hours, at what rate (gallons per minute) must water be pumped into the system?

An important feature of these irrigation systems is their ability to distribute water evenly. We'll return to this problem in Problem 7.8 in Chapter 7 to see what is required to get a uniform water distribution.

3
Mathcad Functions

Optics

The use of *optics* can be dated back at least to the time when someone noticed that putting a curved piece of metal behind a lamp could concentrate the light. Or perhaps it goes back to the time when someone noticed that the campfire felt warmer if it was built next to a large rock. In any case, while the use of optics might be nearly as old as humanity itself, optics is a field that has seen major changes in the past 20 years. And these changes are affecting society in a variety of ways:

- Fiber-optic communication systems are now widely used and will become even more commonplace in the future.
- Lasers have moved from laboratory bench tops to supermarket checkout scanners and CD players.
- Traffic signs are much more visible at night now that they incorporate microreflective optical beads.

As these optical systems continue to develop, they will require the skills of engineers from many disciplines. Fiber-optic communication systems alone will make use of the following kinds of scientists and engineers:

Sections

3.1   Mathcad Functions
3.2   Elementary Mathematics Functions and Operators
3.3   Trigonometric Functions
3.4   Advanced Mathematics Functions
3.5   String Functions
3.6   User-written Functions

Objectives

After reading this chapter, you will

- understand what a function is, and how functions are used in Mathcad
- be able to use Mathcad's built-in functions
- know what functions Mathcad provides for handling logarithms
- know what trigonometric functions are available as built-in functions in Mathcad
- be aware of a few advanced math functions that Mathcad provides for specialized applications
- know that Mathcad provides functions for handling character strings
- know how to write your own functions, when necessary

- *Material scientists*, to select and manipulate the properties of glass.
- *Chemical engineers*, to design glass-processing facilities.
- *Mechanical engineers*, to develop equipment designs for fiber production, cladding, wrapping, handling, and installation.
- *Civil engineers*, to design the support structures for transmission lines.
- *Electrical engineers*, to develop data transfer techniques.
- *Computer scientists*, to design systems to encode and decode transmissions.

Optics is a field that has been studied for centuries, but still has room for growth and continues to change the way we live. Any field that uses technology in society is a field in which engineers will be working.

3.1 MATHCAD FUNCTIONS

In a programming language, the term *function* is used to mean a piece of the program dedicated to a particular calculation. A function accepts input from a list of *parameters*, performs calculations, and then returns a value or a set of values. Functions are used whenever you want to

- perform the same calculations multiple times using different input values,
- reuse the function in another program without retyping it, or
- make a complex program easier to comprehend.

Mathcad's functions work the same way and serve the same purposes. They receive input from a parameter list, perform a calculation, and return a value or a set of values. Mathcad's functions are useful when you need to perform the same calculation multiple times. You can also cut and paste a function from one worksheet to another.

PROFESSIONAL SUCCESS

When do you use a calculator to solve a problem, and when should you use a computer?

These three questions will help you make this decision:

1. Is the calculation long and involved?
2. Will you need to perform the same calculation numerous times?
3. Do you need to document the results for the future, either to give them to someone else or for you own reference?

A "yes" answer to any of these questions suggests that you should consider using a computer. Moreover, a "yes" to the second question suggests that you may want to write a reusable function to solve the problem.

Mathcad provides a wide assortment of built-in functions, as well as allow you to write your own functions. While most Mathcad functions *can* accept arrays as input, some Mathcad functions *require* data arrays as input. For example, the mean ( ) function needs a column of data values in order to compute the average value of the data set. The functions that take arrays as input will be discussed in later chapters, after we describe how Mathcad handles matrices. In this chapter, we present only functions that can take single-valued (scalar) inputs.

The following are the commonly used scalar functions:

- elementary mathematical functions and operators;
- trigonometric functions;

- advanced mathematical functions;
- string functions;

3.2 ELEMENTARY MATHEMATICS FUNCTIONS AND OPERATORS

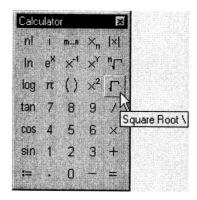

Many of the functions available in programming languages are implemented as *operators* in Mathcad. For example, to take the square root of four in FORTRAN, you would use SQRT(4). SQRT( ) is FORTRAN's square-root function. In Mathcad, the square-root symbol is an operator available on the Calculator Toolbar, and shows up in a Mathcad worksheet just as you would write it on paper. Here are a couple of examples of Mathcad's square-root function:

$$\sqrt{4} = 2$$

$$\sqrt{\frac{4 - \pi}{2}} = 0.655$$

Notice that the square-root symbol changes size as necessary as you enter your equation.

3.2.1 Common Mathematical Operators

The following table presents some common mathematical operators in Mathcad:

TABLE 3.1

OPERATOR	MATHEMATICAL OPERATION	SOURCE PALETTE	ALTERNATE KEYSTROKE
Square Root	$\sqrt{}$	Calculator toolbar	[\] (backslash)
n^{th} Root	$\sqrt[n]{}$	Calculator toolbar	[Ctrl–\]
Absolute Value	$|x|$	Calculator toolbar	[|] (vertical bar)
Factorial	$x!$	Calculator toolbar	[!]
Summation	Σx	Calculator toolbar	[Shift–4]
Product	$\prod x$	Calculus toolbar	[Shift–3]
NOT	$\neg A$	Boolean toolbar	[Ctrl–Shift–1]
AND	$A \wedge B$	Boolean toolbar	[Ctrl–Shift–7]
OR	$A \vee B$	Boolean toolbar	[Ctrl–Shift–6]
XOR	$A \oplus B$	Boolean toolbar	[Ctrl–Shift–5]

Many additional mathematical operations are available as built-in functions. Some of these elementary mathematical functions are the logarithm and exponentiation functions and the round-off and truncation functions.

Note: Mathcad's help files include descriptions of every built-in function, including information about requirements on arguments or parameters. Excerpts from the Mathcad help files are shown in text boxes.[1]

3.2.2 Logarithm and Exponentiation Functions

The following box shows Mathcad's logarithm and exponentiation functions:

```
exp(z)      The number e raised to the power z
log(z,b)    Base b logarithm of z. If b is omitted, base 10 log of z.
ln(z)       Natural logarithm (base e) of z
```

Arguments:
- z must be a scalar (real, complex, or imaginary).
- z must be dimensionless.
- For log and ln functions, z cannot be zero.

But e is also a predefined variable in Mathcad, so the following two calculations are equivalent:

$$\exp(3) = 20.086$$
$$e^3 = 20.086$$

If you know or can guess a function name and want to see the Help for information on the function in order to find out what arguments are required, type the function name on the worksheet, and while the edit lines are still around the function name, press the [F1] key. If Mathcad recognizes the function name you entered, it will display information on that function. For example, if you type log and press [F1], the information shown in the preceding text box will be displayed.

If you don't know the function name, use the Help menu and search the Mathcad index for your subject. For example, you can access the index list from the Help menu by clicking Help/Mathcad Help/Index. A search box will appear. When you type the word "functions", you will receive a lot of information about Mathcad's functions.

PRACTICE!

Try out Mathcad's operators and functions. First try these obvious examples:

a. $\sqrt{4}$
b. $\sqrt[3]{8}$
c. $|-7|$ (This is the absolute-value operator from the Calculator Toolbar.)
d. $3!$
e. $\log(100)$

Then try these less obvious examples and check the results with a calculator:

a. $20!$
b. $\ln(-2)$
c. $\exp(-0.4)$

---

[1] The Mathcad help file text is reprinted here with permission of Mathsoft, Inc.

3.2.3 Using QuickPlots to Visualize Functions

Mathcad has a feature called a *QuickPlot* that produces the graph of a function. For example, to obtain a visual display of the natural logarithm function, you would follow these steps:

1. First, create a graph by either selecting X-Y Graph from the Graph Toolbar or pressing [Shift–2].
2. Enter `ln(x)` in the *y*-axis placeholder. (The *x* can be any unused variable, but you need to use the same variable in step 3.)
3. Enter *x* in the *x*-axis placeholder.

The result will look like this:

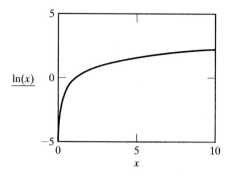

Mathcad shows what the function looks like over an arbitrary range of x values. The default range is −10 to +10, but the natural logarithm is not defined for negative numbers, so Mathcad used only positive numbers for this function. If you want to see the plot over a different range, just click on the *x*-axis and change the plot limits.

3.2.4 3-D QuickPlots

QuickPlots can also be used to visualize functions of two variables. For example, the function $Z(x, y) = 2x^2 - 2y^2$ can be plotted as a contour plot as follows:

1. Define the function in your worksheet.
2. Create the graph by either selecting Surface Graph from the Graph toolbar or pressing [Ctrl-2].
3. Put the variable name, Z, in the placeholder on the surface plot.
4. Double-click on the graph to change display characteristics if desired.

The following is the resulting graph:

3.3 TRIGONOMETRIC FUNCTIONS

Mathcad's trigonometric functions work with angles in radians, but deg is a predefined unit, so you can also display your calculated results in degrees. In addition, there is a predefined unit rad. You probably won't need to use it, since Mathcad assumes *radians* if the unit is not included, but you can include the rad unit label to make your worksheet easier to read. For example, you might have

```
angle := 2·π
angle = 6.283
angle = 6.283°rad
angle = 360°deg
```

Mathcad's basic trigonometric functions are displayed in the following box:

```
sin(z)    cos(z)
tan(z)    cot(z)
sec(z)    csc(z)
```

Arguments:
- z must be in radians.
- z must be a scalar (real, complex, or imaginary).
- z must be dimensionless.

3.3.1 Standard Trigonometric Functions

Mathcad's trigonometric functions require angles in radians, but as the examples that follow illustrate, you can work with either degrees or radians in Mathcad. To understand how this works, remember how Mathcad handles units: When you enter a value with units, Mathcad converts those units to its base units and stores the value. Then it performs the calculation using the base units and displays the result:

```
sin(30·deg) = 0.5
cos (π) = -1
```

When sin(30·deg) = was entered in the preceding example, Mathcad converted the units that were entered (degrees) to base units (radians) and stored the value. Then the sine was computed using the stored value in base units, and the result was displayed. In the second example, no units were entered with the π, so Mathcad used the default unit, radians, and computed the result.

Remember: You can work with degrees with Mathcad's trigonometric functions, but you must include the deg unit abbreviation on each angle. When you do so, Mathcad automatically converts the units to radians before performing the calculation.

PROFESSIONAL SUCCESS

Validate your functions and worksheets as you develop them.

If you are "pretty sure" that the tangent of an angle is equal to the sine of the angle divided by the cosine of the angle, test the tan( ) function before building it into your worksheet. Learning to devise useful tests is a valuable skill. In this case, you could test the relationship between the functions like this:

$$\tan(30 \cdot \text{deg}) = 0.577$$

$$\frac{\sin(30 \cdot \text{deg})}{\cos(30 \cdot \text{deg})} = 0.577$$

It only takes a second to test, and building a number of "pretty sure" items into a worksheet will quickly lead to a result with low confidence.

3.3.2 Inverse Trigonometric Functions

The following box shows Mathcad's inverse trigonometric functions:

```
asin(z)
acos(z)
atan(z)
```

Arguments:
- z must be a scalar.
- z must be dimensionless.

Values returned are angles in radians between 0 and 2π. Values returned are from the principal branch of these functions.

The angles returned by these functions will be in radians. You can display the result in degrees by using the `deg` unit abbreviation with it. You can also append the `rad` unit abbreviation; doing so will not change the displayed angle, but it might help someone else understand your results. For example, you might have

```
asin(0.5) = 0.524
asin(0.5) = 30·deg
asin(0.5) = 0.524·rad
```

APPLICATIONS: RESOLVING FORCES

If one person pulls on a rope connected to a hook imbedded in a floor with a force of 400 N and another person pulls on the same hook with a force of 200 N, what is the total force on the hook?

Answer: Can't tell—there's not enough information.

What's missing in the statement is some indication of the direction of the applied forces, since force is a vector. If two people are pulling in the same direction, then the combined force is 600 N. But if they are pulling in different directions, it's a little tougher to determine the net force on the hook. To help find the answer, we often *resolve* the forces into horizontal and vertical components:

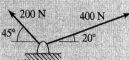

One person pulls to the right on a rope connected to a hook imbedded in a floor with a force of 400 N at an angle of 20° from the horizontal. Another person pulls to the left on the same hook with a force of 200 N at an angle of 45° from the horizontal. What is the net force on the hook?

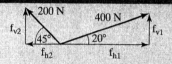

Since both people are pulling upwards, their vertical contributions add. But since one is pulling left and the other right, they are (in part) counteracting each other's efforts. To quantify this distribution of forces, we can calculate the horizontal and vertical components of the force being applied by each person. Mathcad's trigonometric functions are helpful for these calculations.

The 400-N force from person 1 resolves into a vertical component f_{v1} and a horizontal component f_{h1}. The magnitudes of these force components can be calculated as follows:

```
f_v1 := 400·N·sin(20·deg)    f_v1 = 136.808·N
f_h1 := 400·N·cos(20·deg)    f_h1 = 375.877·N
```

Similarly, the 200-N force from person 2 can be resolved into the following component forces:

```
f_v2 := 200·N·sin(45·deg)    f_v2 = 141.421·N
f_h2 := 200·N·cos(45·deg)    f_h2 = 141.421·N
```

Actually, force component f_{h2} would usually be written as $f_{h2} = -141.421$ N, since it is pointed in the $-x$ direction. If all angles had been measured from the same position (usually the three-o'clock angle is called $0°$), the angle on the 200-N force would have been at $135°$ and the signs would have taken care of themselves:

$f_{h2} := 200 \cdot N \cdot \cos(135 \cdot \deg)$ $f_{h2} = -141.421 \circ N$

Once the force components have been computed, the net force in the horizontal and vertical directions can be determined. (Force f_{h2} has a negative value in this calculation.) We obtain

$f_{v\_net} := f_{v1} + f_{v2}$ $f_{v\_net} = 278.229 \circ N$
$f_{h\_net} := f_{h1} + f_{h2}$ $f_{h\_net} = 234.456 \circ N$

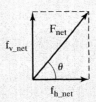

The net horizontal and vertical components can be recombined to find a combined net force on the hook, F_{net}, and angle, θ:

$F_{net} := \sqrt{f_{h\_net}^2 + f_{v\_net}^2}$ $F_{net} = 363.842°N$

$\theta := \mathrm{atan}\left(\dfrac{f_{v\_net}}{f_{h\_net}}\right)$ $\theta = 49.88°$ deg

3.3.3 Hyperbolic Trigonometric Functions

Mathcad's hyperbolic trigonometric functions are as follows

```
sinh(z)    cosh(z)
tanh(z)    csch(z)
sech(z)    coth(z)
```

Arguments:
- z must be in radians.
- z must be a scalar.
- z must be dimensionless.

3.3.4 Inverse Hyperbolic Trigonometric Functions

The following box shows the inverse hyperbolic trigonometric functions in Mathcad:

```
asinh(z)
acosh(z)
atanh(z)
```

Arguments:
- z must be a scalar.
- z must be dimensionless.

Values returned are from the principal branch of these functions.

The angles returned by the inverse hyperbolic trigonometric functions will be in radians, but you can put deg in the units placeholder to convert the displayed units.

PRACTICE!

> Try out these Mathcad trigonometric functions:
>
> a. $\sin(\pi/4)$
> b. $\sin(90 \cdot \deg)$
> c. $\cos(180 \cdot \deg)$
> d. $\operatorname{asin}(0)$
> e. $\operatorname{acos}(2 \cdot \pi)$
>
> Try QuickPlots of these functions over the indicated ranges:
>
> a. $\sin(x) \quad -\pi \leq x \leq \pi$
> b. $\tan(x) \quad -\pi/2 \leq x \leq \pi/2$
> c. $\sinh(x) \quad -10 \leq x \leq 10$

3.4 ADVANCED MATHEMATICS FUNCTIONS

3.4.1 Round-off and Truncation Functions

Mathcad provides two functions that determine the nearest integer value. The `floor(x)` function returns the largest integer that is less than or equal to x, while the `ceil(x)` function returns the smallest integer that is greater than or equal to x. To remove the decimal places following a value, use the `floor()` function. For example, to find the greatest integer less than or equal to π, either of the following will work:

```
floor(3.1416) = 3
floor(π) = 3
```

The second example used Mathcad's predefined variable for π.

If you want the values after the decimal point (i.e., the mantissa), use the `mod()` function, which performs *modulo division*, returning the remainder after the division calculation:

`mod(x,y)`   This function returns the remainder on dividing x by y. The result has the same sign as x.

Arguments:
- x and y should both be real scalars.
- y must be nonzero.

Using the example of π again, we have

```
mod(3.1416,3) = 0.1416
mod(π,3) = 0.1416
```

Notice that the `mod()` function will return the trailing decimal places only when the y argument is the integer less than or equal to x. You can create a user-written function that will always return the trailing decimal places using `mod()` and `floor()` together like this:

```
trail(x) := mod(x,floor(x))
trail(3.1416) = 0.1416
trail(π) = 0.1416
```

The new `trail( )` function doesn't work correctly when x is negative, because it tries to divide −3.1416 by −4, which does not yield the trailing digits. An improved version uses Mathcad's absolute-value operator, which is available on the Calculator Toolbar. For example, we might have

```
trail(x) := mod(x, floor(|x|))
trail(3.1416) = 0.1416
trail(-3.1416) = -0.1416
```

The result has the same sign as the x value, which might be useful. If not, another absolute-value operator around the entire right-hand side would force a positive result:

```
trail(x) := |mod(x, floor(|x|))|
trail(3.1416) = 0.1416
trail(-3.1416) = 0.1416
```

PRACTICE!

You can compute the mantissa by subtracting `floor(x)` from x, but this approach also has trouble with negative numbers:

```
mantissa(x) := x-floor(x)
mantissa(3.1416) = 0.1416
mantissa(-3.1416) = 0.8584
```

Modify the `mantissa( )` function shown here so that it returns the correct mantissa when x is negative.

3.4.2 The if( ) Function

Mathcad provides an `if( )` function for making logical decisions, as well as functions for handling discontinuous systems and generating random numbers.

Rules for the `if( )` function in Mathcad are as follows:

`if(cond, Tval, Fval)`   This function returns one of two values, depending on the value of a logical condition.

Arguments:
- cond is usually an expression involving a logical operator. For example, you can use

  (i < 2)
  (x < 1) * (x > 0)   for an "and" gate
  (x > 1) + (x < 0)   for an "or" gate

- Tval is the value returned when cond is true.
- Fval is the value returned when cond is false.

The value that is returned can be a simple scalar value, a computed value, or even a text string. For example, the `if( )` function can be used to determine whether water would be a liquid or a solid by checking the temperature of the water:

```
state(Temp) := if(Temp > 0, "liquid", "solid")
state(25) = "liquid"
state(-3) = "solid"
```

Tval and Fval can be formulas. Thus, we can create an absolute-value function by using if ( ) and changing the sign on x if $x < 0$:

```
abs(x) := if(x ≥ 0, x, -x)
abs(12) = 12
abs(-6) = 6
```

PRACTICE

> Write an expression that returns a value of 1 if the volume in a tank is greater than or equal to 5,000 gallons and a value of 0 otherwise. An expression like this might be used to control a valve through which a liquid flows into the tank, shutting the valve when the tank is full. Test your expression in Mathcad.

3.4.3 Discontinuous Functions

Mathcad supports the following discontinuous functions:

KRONECKER DELTA

δ(m,n) This function returns 1 if m = n and returns 0 otherwise.

Arguments:
- Both m and n must be integers without units.

HEAVISIDE STEP FUNCTION

ϕ (x) This function returns 0 if x is negative and returns 1 otherwise.

Arguments:
- x must be a real scalar.

3.4.4 Boolean (Logical) Operators

Mathcad provides the following Boolean (logical) operators on the Boolean toolbar.

- ¬A   NOT Operator–returns 1 if A is zero; returns 0 otherwise.
- A∧B   AND Operator–returns 1 if A and B both have nonzero values; otherwise returns 0.
- A∨B   OR Operator–returns 1 if A or B is nonzero; otherwise returns 0.
- A⊕B   XOR (Exclusive OR) Operator–returns 1 if A or B, but not both, is nonzero; otherwise returns 0.

3.4.5 Random-Number-Generating Function

rnd(x)   Returns a random number (not an integer) between 0 and x.

A set of random numbers generated by rnd(x) will have a uniform distribution. Mathcad provides a number of other random-number-generating functions that produce sets of numbers with various distributions. For example, the rnorm(m, μ, σ) function produces a vector of m values that are normally distributed about the value μ with a standard deviation σ.

3.5 STRING FUNCTIONS

Because text regions can be used anywhere on a worksheet, the need for string-handling functions is greatly reduced in Mathcad. Still, Mathcad does provide the typical functions for manipulating text strings:

`concat(S1,S2)`	Concatenates string `S2` to the end of string `S1`. Returns a string.
`strlen(S)`	Determines the number of characters in the string `S`. Returns an integer.
`substr(S,n,m)`	Extracts a substring of `S`, starting with the character in position n and having at most m characters. The arguments m and n must be integers.
`search(S,SubS,x)`	Finds the starting position of the substring `SubS` in `S`, beginning from position x in `S`.
`str2num(S)`	Converts a string of numbers `S` into a constant.
`num2str(x)`	Converts the number x into a string.

The example that follows illustrates the use of the `search( )` function to find the word `Mathcad` in the string called `MyString`. The position of the `M` in `Mathcad` is returned by the function. Since Mathcad says the first character of the string (the `T` in `This`) is in position zero, the `M` in `Mathcad` is in position 10 (the 11th character of the string). The Mathcad code is

```
MyString := "This is a Mathcad example."
pos := search(MyString, "Mathcad", 0)
pos = 10
pos := search(MyString, "mathcad", 0)
pos = -1
```

The last two lines illustrate that Mathcad differentiates between uppercase and lowercase letters. That is, Mathcad is *case sensitive*. The `search( )` function returns a -1 when the search string is not found.

3.6 USER-WRITTEN FUNCTIONS

When Mathcad does not provide a built-in function, you can always write your own. The ability to use Mathcad's functions inside your own function definitions greatly increases the power of Mathcad to solve engineering problems and makes writing functions much simpler. Several user-written functions have been utilized in this chapter, including the following:

```
trail(x)    := |mod(x,floor(|x|))|
mantissa(x) := x-floor(x)
state(Temp) := if(Temp > 0, "liquid", "solid")
abs(x)      := if(x ≥ 0, x, -x)
```

Each of these user-written functions has a name (e.g., `trail`) and a parameter list on the left, and a mathematical expression of some sort on the right side. In these examples, only a single argument was used in each parameter list, but multiple arguments are allowed and are very common. For example, a function to calculate the surface area of a cylinder could be written as

$$A_{cyl}(r,L) := 2 \cdot \pi \cdot r^2 + 2 \cdot \pi \cdot r \cdot L$$

Once the function has been defined, it can be used multiple times in the worksheet. In the following example, the function is employed four times, illustrating how functions can be used with and without units and with both values and variables:

$$A_{cyl}(7, 21) = 1.232 \cdot 10^3$$
$$A_{cyl}(7 \cdot cm, 21 \cdot cm) = 0.123 \cdot m^2$$
$$A_{cyl}(7 \cdot cm, 21 \cdot cm) = 1.232 \cdot 10^3 \circ cm^2$$
$$R := 7 \cdot cm \quad L := 21 \cdot cm$$
$$A_{cyl}(R, L) = 1.232 \cdot 10^3 \circ cm^2$$

Recall that the A_{cyl} function was defined using r and L for arguments. In a function definition, the arguments are dummy variables. That is, they show what argument is used in which location in the calculation, but neither R nor L needs to have defined values when the function is declared. They are just placeholders. When the function is used, the values in the R and L placeholders are put into the appropriate spots in the equation, and the computed area is returned.

In a function, you can use arguments with or without units, and the arguments can be values or defined variables. In the last calculation, R and L were used for the radius and length of the cylinder, respectively. This choice of variable names is reasonable, but entirely irrelevant to Mathcad. Variables like Q and Z would work equally well, but they would make the worksheet harder to understand. For example, we might have

$$Q := 7 \cdot cm$$
$$Z := 21 \cdot cm$$
$$A_{cyl}(Q, Z) = 1.232 \cdot 10^3 \circ cm^2$$

PRACTICE!

Define a function for computing the volume of a rectangular box, given the height, width, and length of the box. Then use the function to compute the volumes of several boxes:

a. H = 1, W = 1, L = 1
b. H = 1, W = 1, L = 10
c. H = 1 cm, W = 1 cm, L = 10 cm
d. H = 1 cm, W = 1 m, L = 10 ft

APPLICATIONS: FIBER OPTICS

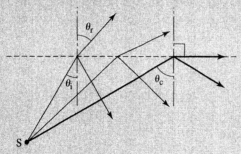

According to the *law of refraction*, the angle of incidence, θ_i, is related to the angle of refraction, θ_r, by the *index of refraction* n, for the two materials through which light passes. That is,

$$n_1 \sin(\theta_r) = n_2 \sin(\theta_2).$$

For light in a glass fiber ($n_1 \approx 1.5$) reflecting and refracting at the air ($n_2 \approx 1$) interface with an angle of incidence of 20°, what is the angle of refraction?

To solve this problem, we will use Mathcad's sin() and asin() functions and the built-in unit deg:

$$\theta_r := asin\left(\frac{1.5 \cdot \sin(20 \cdot deg)}{1}\right)$$

$$\theta_r = 30.866 \circ deg$$

When the angle of reflection is 90°, the angle of incidence is called the *critical angle*. For the fiber-optic

material in air, what is the critical angle?
The critical angle is

$$\theta_r := \operatorname{asin}\left(\frac{1.5 \cdot \sin(90 \cdot \deg)}{1.5}\right)$$

$$\theta_c = 41.81 \circ \deg$$

Note: The "1" in the first equation is there just to show how the index of refraction of air enters into the equation. It could be left out with no impact on the result.

The critical angle is important because, if you keep the angle of incidence greater than the critical angle, you get total internal reflection—there is no refracted ray, and thus no light is lost because of refraction. This is extremely important if the fiber-optic material is to be used for transmitting signals, since refracted light shortens the potential transmission distance.

FIBER CLADDING

If the medium outside the fiber is something other than air, the critical angle changes. For example, if the fiber is in water ($n_2 \approx 1.33$), the critical angle is higher:

$$\theta_r := \operatorname{asin}\left(\frac{1.33 \cdot \sin(90 \cdot \deg)}{1.5}\right)$$

$$\theta_c = 62.457 \circ \deg$$

By increasing the refractive index of the material on the outside of the fiber, the angle of incidence can be closer to perpendicular to the wall of the fiber and still have total internal reflection. If the material outside the fiber has a refractive index greater than 1.5, you get total internal reflection for *any* angle of incidence. To ensure that this is the case, fiber-optic materials are often coated with a second layer of glass with a slightly higher refractive index. This process is called *cladding* the fibers.

SUMMARY

In this chapter, you learned that a function is a reusable equation that accepts arguments, uses the argument values in a calculation, and returns the result(s). Once a function has been defined in a worksheet, it can be used multiple times in that worksheet. You have learned to write your own functions and to use Mathcad's built-in functions in the following areas:

MATHCAD SUMMARY

Logarithm and Exponentiation

```
exp(x)      Raises e to the power x.
ln(x)       Returns the natural logarithm of x.
log(x)      Returns the base-10 logarithm of x.
```

Trigonometric Functions The units on the angle a in these functions must be radians. If you want to work in degrees, replace a by a·deg, where deg is a predefined unit for degrees:

```
sin(a)      Returns the sine of a.
cos(a)      Returns the cosine of a.
tan(a)      Returns the tangent of a.
sec(a)      Returns the secant of a.
csc(a)      Returns the cosecant of a.
cot(a)      Returns the cotangent of a.
```

Inverse Trigonometric Functions These functions take a value between −1 and 1 and return the angle in radians. You can display the result in degrees by using deg.

```
asin(x)     Arcsine—returns the angle that has a sine value equal to x.
acos(x)     Arccosine—returns the angle that has a cosine value equal to x.
atan(x)     Arctangent—returns the angle that has a tangent value equal to x.
```

Hyperbolic Trigonometric Functions
    sinh(a)   cosh(a)
    tanh(a)   csch(a)
    sech(a)   coth(a)

Inverse Hyperbolic Trigonometric Functions
    asinh(x)
    acosh(x)
    atanh(x)

Round-Off and Truncation Functions
    floor(x)   Returns the largest integer that is less than or equal to x.
    ceil(x)   Returns the smallest integer that is greater than or equal to x.
    mod(x,y)   Returns the remainder after dividing x by y.

Advanced Mathematics Functions
    if(cond,Tval,Fval)   Returns Tval or Fval, depending on the value of the logical condition cond.
    δ(m,n)   Kronecker delta—returns 1 if m = n and returns 0 otherwise.
    ϕ(x)   Heaviside step function—returns 0 if x is negative and returns 1 otherwise.
    rnd(x)   Returns a random number (not an integer) between 0 and x.

Boolean (Logical) Operators
    ¬A   NOT Operator—returns 1 if A is zero; returns 0 otherwise.
    A∧B   AND Operator—returns 1 if A and B both have nonzero values; otherwise returns 0.
    A∨B   OR Operator—returns 1 if A or B is nonzero; otherwise returns 0.
    A⊕B   XOR (Exclusive OR) Operator—returns 1 if A or B, but not both, is nonzero; otherwise returns 0.

String Functions
    concat(S1,S2)   Concatenates string S2 to the end of string S1. Returns a string.
    strlen(S)   Determines the number of characters in the string S. Returns an integer.
    substr(S,n,m)   Extracts a substring of S, starting with the character in position n and having at most m characters. The arguments m and n must be integers.
    search(S,SubS,x) Finds the starting position of the substring SubS in S, beginning from position x in S.
    str2num(S)   Converts a string of numbers S into a constant.
    num2str(x)   Converts the number x into a string.

KEY TERMS

boolean (logical) operator   character string   exponentiation
case sensitive   discontinuous function   function

logarithm	radian	string function
modulo division	random number	toolbar
operator	return value	trigonometric
parameter	round-off	truncate

Problems

1. **FORCE AND PRESSURE**

   A force of 800 N is applied to a piston with a surface area of 24 cm².

   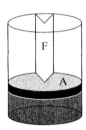

   a. What pressure (Pa) is applied to the fluid? Use

   $$P = \frac{F}{A}$$

   b. What is the diameter (cm) of a piston with a (circular) surface area of 24 cm²?

2. **CALCULATING BUBBLE SIZE**

   Very small gas bubbles (e.g., air bubbles less than 1 mm in diameter in water) rising through liquids are roughly spherical in shape.

   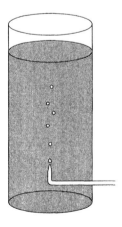

   a. What volume of air (mm³) is contained in a spherical bubble 1 mm in diameter?

   b. What is the surface area (mm²) of a spherical bubble 1 mm in diameter?

c. Calculate the diameter of a spherical bubble containing 1.2×10^{-7} mole of nitrogen at 25°C at 1 atm. Use the formulas

$$V = \frac{4}{3}\pi R^3$$

and

$$A = 4\pi R^3.$$

3. RELATIONSHIPS BETWEEN TRIGONOMETRIC FUNCTIONS

Devise a test to demonstrate the validity of the following common trigonometric formulas:

a. $\sin(A + B) = \sin(A) \cdot \cos(B) + \cos(A) \cdot \sin(B)$
b. $\sin(2 \cdot A) = 2 \cdot \sin(A) \cdot \cos(A)$
c. $\sin^2(A) = 1/2 - 1/2 \cdot \cos(2 \cdot A)$

What values of A and B should be used to thoroughly test these functions?

Note: In Mathcad, `sin`^{2}`(A)` should be entered as `sin(A)`^{2}. This causes `sin(A)` to be evaluated first and that result to be squared.

4. CALCULATING ACREAGE

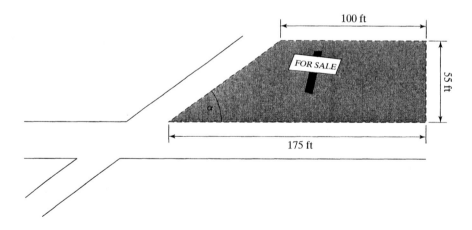

An odd-shaped corner lot is up for sale. The "going rate" for property in the area is $3.60 per square foot.

a. Determine the corner angle, α (in degrees).
b. What is the area of the lot in square feet? In acres?
c. How much should the seller ask for the property?

5. USING THE ANGLE OF REFRACTION TO MEASURE THE INDEX OF REFRACTION

A laser in air ($n_2 \approx 1$) is aimed at the surface of a liquid with an angle of incidence of 45°. A photodetector is moved through the liquid until the beam is located at an angle of refraction of 32°. What is the index of refraction of the liquid?

6. **MEASURING HEIGHTS**

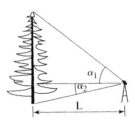

A surveyor's *transit* is a telescopelike device mounted on a protractor that allows the surveyor to sight distant points and then read the angle of the scope off the scale. With a transit it is possible to get accurate measurements of large objects.

In this problem, a transit is being used to measure the height of a tree. The transit is set up 80 feet from the base of the tree (L = 80 ft), and the sights are set on the top of the tree. The angle reads 43.53° from the horizontal. Then the sights are set at the base of the tree, and the second angle is found to be 3.58°. Given this information, how tall is the tree?

7. **LADDER SAFETY**

Ladders are frequently used tools, but they are also the source of many injuries. Improperly positioning a ladder is a common cause of falls.

Ladders are supposed to be set against walls at an angle such that the base of the ladder stands away from the wall 1 foot for every 4 feet of working length of the ladder. So the base of a 12-foot ladder with the top resting against a wall (as shown in the foregoing figure) should be 3 feet away from the wall. At this angle, most people can stand up straight on a rung and reach out and easily grab another rung with their hands. The 1-in-4 rule is for comfort and for safety—and it is written into OSHA regulations (OSHA Standards, 29.b.5.i).

   a. What is the angle θ between the floor and the ladder if the 1-in-4 rule is used?

   b. If the 1-in-4 rule is used, at what height does the top of a 12-foot ladder touch the wall?

8. SOLVING QUADRATIC EQUATIONS

A quadratic equation can be written in the form

$$ax^2 + bx + c = 0 \quad (a \neq 0).$$

These equations have two solutions, which can be found by using the formula

$$x = \frac{-b \pm \sqrt{b^2 - 4ac}}{2a}$$

Find the solutions to the following equations:

a. $-2x^2 + 3x + 4 = 0.$ (Expect real roots, since $b^2 - 4ac > 0$.)
b. $-4x^2 + 4x - 1 = 0.$ (Expect real, equal roots, since $b^2 - 4ac = 0$.)
c. $3x^2 + 2x + 1 = 0.$ (Expect imaginary roots, since $b^2 - 4ac < 0$.)

9. HEATING A HOT TUB

A 500-gallon hot tub has been unused for some time, and the water in the tub is at 65°F.

Hot water (130°F) will be added to the tub at a rate of 5 gallons per minute. The temperature in the hot tub can be calculated with the equation

$$T = T_{IN} - (T_{IN} - T_{START})e^{-\frac{Q}{V}t},$$

where

T	is the temperature of the water already in the tub,
T_{IN}	is the temperature of the water flowing into the tub (130°F),
T_{START}	is the initial temperature of the water in the hot tub (65°F),
Q	is the hot-water flow rate (5 gpm)
V	is the volume of the tub (500 gallons), and
T	is the elapsed time since the hot water started flowing.

a. What will be the temperature of the water in the hot tub after one hour?
b. How long will it take the water in the tub to reach a temperature of 110°F?
c. If the temperature of the water flowing into the tub could be raised to 150°F, how long would it take to warm the water to 110°F?

10. VAPOR PRESSURE

Knowing the vapor pressure is important in designing equipment that will or might contain boiling or highly volatile liquids. When the pressure in a vessel is at or below the liquid's vapor pressure, the liquid will boil. If you are designing a boiler, then designing it to operate at the liquid's vapor pressure is a pretty good idea. But if you are designing a waste solvent storage facility, you want the pressure to be significantly higher than the liquid's vapor pressure. In either case, you need to be able to calculate the liquid's vapor pressure.

Antoine's equation is a common way to calculate vapor pressures for common liquids. It requires three coefficients that are unique to each liquid. If the coefficients are available, Antoine's equation[2] is easy to use. The equation is

$$\log(p_{vapor}) = A - \frac{B}{T + C},$$

where

- p_{vapor} is the vapor pressure of the liquid in mm Hg,
- T is the temperature of the liquid in °C, and
- A, B, C are coefficients for the particular liquid (found in tables).

In using Antoine's equation, keep the following important considerations in mind:

1. Antoine's equation is a *dimensional equation;* that is, specific units must be used to get correct results.
2. The log() function is not supposed to be used on a value with units, but that's the way Antoine's equation was written. Mathcad, however, will not allow you to take the logarithm of a value with units, so you must work without units while using Antoine's equation.
3. There are numerous forms of Antoine's equation, each requiring specific units. The coefficients are always designed to go along with a particular version of the equation. You cannot use Antoine coefficients from one text in a version of Antoine's equation taken from a different text.

TEST ANTOINE'S EQUATION

Water boils at 100°C at 1 atm pressure. So the vapor pressure of water at 100°C should be 1 atm, or 760 mm Hg. Check this using Antoine's equation. For water,

$$A = 7.96671 \quad B = 1668.21 \quad C = 228.0$$

USE ANTOINE'S EQUATION

a. A chemistry laboratory stores acetone at a typical room temperature of 25°C (77°F) with no problems. What would happen if, on a hot summer day, the air-conditioning failed and a storage cabinet warmed by the sun reached 50°C (122°F)? Would the stored acetone boil?

b. At what temperature would the acetone boil at a pressure of 1 atm? For acetone:

$$A = 7.02447 \quad B = 1161.0 \quad C = 224$$

---

[2] This version of Antoine's equation is from *Elementary Principles of Chemical Processes*, by R. M. Felder and R. W. Rousseau (New York: Wiley, 1978).

11. CALCULATING THE VOLUME AND MASS OF A SUBSTANCE IN A STORAGE TANK

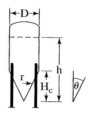

Vertical tanks with conical base sections are commonly used for storing solid materials such as grain, gravel, salt, catalyst particles, etc. The sloping sides help prevent plugging as the material is withdrawn.

If the height of solid in the tank is less than the height of the conical section, H_c, then the volume is computed by using the formula for the volume of a cone:

$$V = \frac{1}{3}\pi r^2 h$$

But if the tank is filled to a depth greater than H_c, then the volume is the sum of the filled conical section, V_{cone}, and a cylindrical section, V_{cyl}, of height $h-H_c$:

$$V = V_{cone} + V_{cyl}$$
$$V = \frac{1}{3}\pi R^2 H_C + \pi R^2 (h - H_C)$$

Here, R = D/2.

The radius of the tank depends on h if $h < H_c$, but has a value of R when $h = H_c$. The if( ) function can be used to return the correct value of r for any h:

$$r(h) := \text{if } (h < H_c, h \cdot \tan(\theta), R)$$

where θ is the angle of the sloping walls. (Variables H, R, and θ must be defined before using this function.)

When working with solids, you often want to know the mass as well as the volume. The mass can be determined from the volume using the *apparent density*, ρ_A, which is the mass of a known volume of the granular material (including the air in between the particles) divided by the known volume. Since the air between the particles is much less dense than the solid particles themselves, the apparent density of a granular solid is typically much smaller than the true density of any individual particle.

a. Use Mathcad's if( ) function to create a function that will return the volume of solids for any height.
b. Use the function from part a to determine the volume and mass of stored material (ρ_A = 20 lb/ft³) in a tank (D = 12 ft, θ = 30°) when the tank is filled to a depth h = 21 ft.

(See related problem in Chapter 5.)

12. **FORCE COMPONENTS AND TENSION IN WIRES**

A 150-kg mass is suspended by wires from two hooks. The lengths of the wires have been adjusted so that the wires are each 50° from horizontal. Assume that the mass of the wires is negligible.

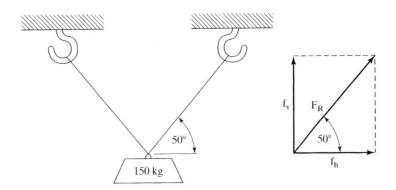

a. Since two hooks support the mass equally, the vertical component of force exerted by either hook will be equal to the force resulting from 75 kg being acted on by gravity. Calculate this vertical component of force, f_v, on the right hook. Express your result in newtons.

b. Compute the horizontal component of force, f_h, by using the result obtained in problem a. and trigonometry.

c. Determine the force exerted on the mass in the direction of the wire F_R, (equal to the tension in the wire).

d. If you moved the hooks farther apart to reduce the angle from 50° to 30°, would the tension in the wires increase or decrease? Why?

13. **MULTIPLE LOADS**

If two 150-kg masses are suspended on a wire, such that the section between the loads (wire B) is horizontal, then wire B is under tension, but is doing no lifting. The entire weight of the 150-kg mass on the right is being held up by the vertical component of the force in wire C. In the same way, the mass on the left is being supported entirely by the vertical component of the force in wire A.

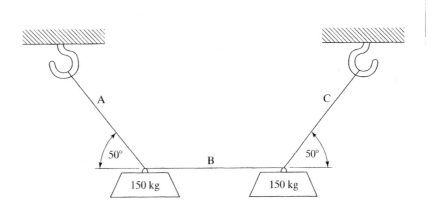

Calculate

    a. the vertical force component in wires A and C.

    b. the horizontal force component in each wire.

    c. the tension in wire A.

14.   TENSION AND ANGLES

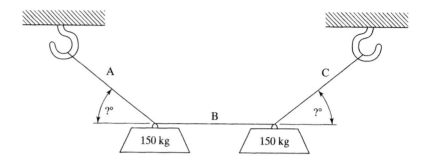

If the hooks are pulled farther apart, the tension in wire B will change, and the angle of wires A and C with respect to the horizontal will also change.

If the hooks are pulled apart until the tension in wire B is 2,000 N, determine

    a. the angle between the horizontal and wire C, and

    b. the tension in wire C.

How does the angle in part a change if the tension in wire B is increased to 3,000 N?

15.   DESIGNING A SUSPENSION BRIDGE

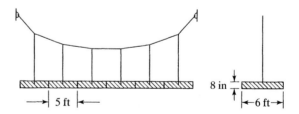

After fighting some particularly snarled traffic to get from their apartment to the pizza place across the street, some engineers decided that they should build a suspension bridge from one side of the street to the other. On the back side of a napkin, they sketched out the initial design shown in the accompanying figure. The plan calls for a single cable to be hung between the buildings, with one support wire to each deck section. The deck is to be 6 feet wide to allow pedestrian traffic on either side of the support wires. The bridge will be 30 feet long and made up of six 5-foot sections. The engineers want their design to be environmentally friendly, so the design calls for the decking to be made of a material constructed of used tires, with a density of 88 lb/ft^3.

Note: To help keep things straight, we use the term "wire" for the vertical wires connected to the deck sections and the term "cable" for the main suspension between the buildings.

a. What is the mass of each deck section?
b. What force is a deck section imparting to its suspension wire?

The engineers had quite a discussion about how much tension to impose on the bridge. One wanted to minimize the stress on the buildings by keeping the tension low, but others thought that would produce a "droopy" bridge. They finally decided that the tension in the center (horizontal) section would be five times the force that any deck section put on its suspension wire.

c. What is the horizontal component of force in any of the cable sections?
d. Calculate all the forces and angles in the main cable segments.

Hint: Since the center segment of the cable has only a horizontal force component and you can calculate the tension in that segment, start in the center and work your way towards the outside. Also, since the bridge is symmetrical, you only need to solve for the angles and forces in one-half of the bridge.

At this point, the waiter walked up and made a comment that upset all of their plans. The result was a lively discussion that went on into the night, but no follow-through on the original design.

SO WHAT DID THE WAITER SAY?

"As soon as someone stands on that bridge, the deck will tip sideways and dump the person off."

The engineers spent the rest of the evening designing solutions to this problem. Do you have any ideas?

4
Working with Matrices

TOTAL RECYCLE

NASA's Mission to Mars projects have focused attention on the idea of recovering and reusing oxygen, water, and essential nutrients on extended human excursions into space. Because the quantities of these materials that can be carried along are quite limited, the idea is to recycle everything. This concept is called *total recycle*. It is a great idea and an enormous engineering challenge.

The goal for NASA's Advanced Life Support Program[1] is to completely recycle water, oxygen, and food products—but not energy. Energy will be used to fuel the recycling systems. To date, water and oxygen recycle systems have been tested for up to 91 days.[2] A long-term test of a complete recycle system for food products is still many years away

One of the many challenges of designing a total-recycle system is handling the accumulation of chemicals generated at minute levels over long periods of time. For example, traces of copper (from pipes or cooling systems) are not a serious problem, but if the copper should somehow accumulate in the water system, the concentration could increase. The microbes used to degrade organic wastes and

SECTIONS

4.1 Mathcad's Matrix Definitions
4.2 Initializing an Array
4.3 Modifying an Array
4.4 Array Operations
4.5 Array Functions

OBJECTIVES

After reading this chapter, you will

- understand how Mathcad uses the terms "array," "matrix," and "vector"
- know several ways to create a matrix and fill it with values
- be able to add and subtract matrices using Mathcad
- be able to use Mathcad to multiply matrices
- be able to perform element-by-element multiplication of matrices with Mathcad
- be able to use Mathcad to transpose matrices
- be able to invert matrices, if possible, using Mathcad
- be able to find the determinant of a matrix to determine whether inversion is possible
- be aware of Mathcad's built-in functions to manipulate matrices

---

[1]Check NASA's website, <http://advlifesupport.jsc.nasa.gov/>, for more information on the processing steps required to implement the recycle systems.
[2]The Lunar-Mars Life Support Test Project Phase III (90-Day Human Test) was completed in December 1997.

the plants used to consume CO_2 and generate O_2 are both sensitive to copper. The loss of either of these systems would be catastrophic to the Mars mission. Special designs for handling trace contaminants and long-term testing are necessary to be sure that the total-recycle systems will work for extended missions into space.

The practical use of total-recycle concepts is not limited to the Mission to Mars. Just as individuals and communities are becoming increasingly involved in recycling activities, companies are also working to reduce the generation of waste products by recycling and reusing materials. There is a simultaneous push towards substituting less hazardous materials wherever possible. The separation processes required to effectively recycle materials and the equipment changes needed to allow the use of less hazardous materials will provide engineering opportunities for many years to come.

As with NASA's system, as our manufacturing facilities move towards total recycle, we may see more energy being used to fuel the separation systems that allow the materials to be recycled. During a time when there are concerns over the use of fossil fuels, using increasing amounts of energy to reduce wastes is problematic. The search for low-energy processes that simultaneously reduce waste products presents a tremendous challenge for the next generation of engineers and scientists.

4.1 MATHCAD'S MATRIX DEFINITIONS

A *matrix* is a collection of numbers, called *elements*, that are related in some way. We commonly use matrices to hold data sets. For example, if you recorded the temperature of the concrete in a structure over time as the concrete set, the time and temperature values would form a data set of related numbers. This data set would be stored in a computer as a matrix.

Common usage in mathematics calls a single column or row of values a *vector*. If the temperature and time values were stored separately, we would have both a time vector and a temperature vector. A *matrix* is a collection of one or more vectors. That is, a matrix containing a single row or column would also be a vector. On the other hand, a matrix containing three rows and two columns would be called a 3×2 matrix, not a vector. (But you could say the matrix is made up of three row vectors or two column vectors.) So every vector is a matrix, but only single-row and single-column matrices are called vectors.

Mathcad modifies these definitions slightly by adding the term *array*. In Mathcad, a vector has only one row or one column, and a matrix always has at least two rows or two columns. That is, there is no overlap in Mathcad's definitions of vectors and matrices. Mathcad uses the term *array* to mean a collection of related values that could be either a vector or a matrix. Mathcad uses the new term to indicate what type of parameter must be sent to functions that operate on vectors and matrices. For example, you can send either a vector or a matrix to the rows(A) function, so Mathcad's help files show an A as the function's parameter to indicate that either a vector or a matrix is acceptable.

DEFINITIONS USED IN Mathcad HELP FILES:

A   Array argument—either a matrix or a vector.
M   Matrix argument—an array with two or more rows or columns.
v   Vector argument—an array containing a single row or column.

The length(v) function requires a vector and will not accept multiple rows or columns. Mathcad indicates this vector-only restriction in its help files by showing a v as the length function's parameter. The determinant operator, |M|, requires a square matrix and will not work on a vector. To show that a matrix is required, Mathcad's help files display an M as the operator's parameter.

Array Origin

By default, Mathcad refers to the first element in a vector as element zero. For a two-dimensional matrix, the *array origin* is 0,0 by default. If you would rather have Mathcad start counting elements at another *index* value, you can change the default by using the ORIGIN variable. The following example illustrates how changing the origin from 0,0 to 1,1 changes the row and column numbering of the array elements:

$$\text{MyArray} := \begin{bmatrix} 12 & 15 & 17 \\ 23 & 25 & 29 \end{bmatrix}$$

$\text{MyArray}_{0,0} = 12 \quad \text{MyArray}_{0,1} = 15$
ORIGIN := 1

$$\text{MyArray} := \begin{bmatrix} 12 & 15 & 17 \\ 23 & 25 & 29 \end{bmatrix}$$

$\text{MyArray}_{1,1} = 12 \quad \text{MyArray}_{1,2} = 15$

Notice that the arrays look the same, but that the array origin has been changed to 1,1 *after* the ORIGIN statement, so the top-left element is now element 1,1 and not 0,0. You can also change the array origin for the entire worksheet. Use the Tools menu, as Tools/Worksheet options, and set the new value for the array origin on the Options dialog box. The origins for all arrays on the worksheet will be changed.

Maximum Array Size

There are two constraints on array sizes in Mathcad:

- If you enter arrays from the keyboard, the arrays may have no more than 100 elements. You can type in multiple arrays and join them to work around this limitation.
- The total number of elements in all arrays is dependent on the amount of memory in your computer, but will always be less than 8×10^6.

4.2 INITIALIZING AN ARRAY

Before an array can be used, it must be filled with values, or initialized. There are a number of methods for initializing an array in Mathcad. You can

- type in the values from the keyboard,
- read the values from a file,
- use an input table to fill the array,
- compute the values by using a function or a range variable, or
- copy and paste the values from another Windows program.

Each of these methods will be discussed in turn.

Method 1: Typing Values into an Array

You can enter a value into a particular element of an array with the use of a standard Mathcad definition. For example, typing [G] [[ ] [3] [,] [2] [:] [6] [4] puts the value 64 in the element at position 3,2 in the G matrix. If the G matrix has not been previously defined, then Mathcad will create a matrix large enough to a have an element at position 3,2 and will fill the undefined elements with zeroes. In the following example, notice that the array origin is 0,0, so the 64 in element 3,2 is in the fourth row and third column:

$$G_{3,2} := 64$$

$$G := \begin{bmatrix} 0 & 0 & 0 \\ 0 & 0 & 0 \\ 0 & 0 & 0 \\ 0 & 0 & 64 \end{bmatrix}$$

Filling a matrix by defining the contents of each element would be extremely slow. Fortunately, Mathcad has provided a better way. Rather than forcing you to define each element individually, Mathcad allows you to define an entire array at one time by first creating an array of placeholders and then filling in the placeholders by typing values from the keyboard. Begin by choosing a variable name for the matrix. We will use G again and create the left side of a definition by using [G] [:]:

$$G := \blacksquare$$

To create the array, select (click on) the placeholder on the right side of the definition, and bring up the Insert Matrix dialog box by (a) choosing Matrix from the Insert menu, (b) choosing Matrix from the Matrix Toolbox, or (c) using the keyboard shortcut [Ctrl-M]. The following figure shows what the dialog box looks like:

Enter the number of rows and columns you want in the G matrix:

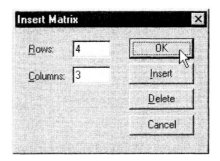

Then click the OK button to insert the matrix of placeholders, and close the dialog box. (The Insert button creates the array of placeholders and leaves the dialog box open, which is useful if you are defining several arrays.) The matrix of placeholders looks like this:

$$G := \begin{bmatrix} \blacksquare & \blacksquare & \blacksquare \\ \blacksquare & \blacksquare & \blacksquare \\ \blacksquare & \blacksquare & \blacksquare \\ \blacksquare & \blacksquare & \blacksquare \end{bmatrix}$$

To complete the definition of array G, simply click on a placeholder and enter the element's value. You can move to another placeholder either by using the mouse or by pressing [Tab] after each entry. The resulting matrix should look like this:

$$G := \begin{bmatrix} 1 & 1 & 1 \\ 2 & 4 & 8 \\ 3 & 9 & 27 \\ 4 & 16 & 64 \end{bmatrix}$$

PRACTICE!

Use the Insert Matrix dialog to create the following matrices:

$$t := (2 \ 4 \ 6 \ 8 \ 10)$$

$$W := \begin{bmatrix} 1 & 2 & 3 & 4 \\ 3 & 1 & 5 & 7 \end{bmatrix}$$

$$y := \begin{bmatrix} 1 \cdot \sec & 2 \cdot \min & 3 \cdot hr \\ 4 \cdot \min & 5 \cdot \min & 6 \cdot \min \end{bmatrix}$$

Method 2: Reading Values from a File

Text, or ASCII, files are commonly used to move data between programs. For example, a recording instrument or data acquisition program might collect data on the concentration of certain substances over time and save the data to a disk as a text file. You could then import the text file into Mathcad as a matrix for further analysis by using the READPRN() function.

A data file on a disk can be read and assigned to a Mathcad array variable by building the read operation into the matrix definition, like this:

```
C := READPRN("A:MyData.txt")
```

C =

	0	1
0	0	50
1	10	48.2
2	20	46.5
3	30	44.8
4	40	43.2
5	50	41.6
6	60	40.1
7	70	38.7
8	80	37.3
9	90	35.9

The text string sent to the READPRN( ) function tells Mathcad to read the MyData.txt file found on drive A:. This data file contains 25 rows, only part of which are displayed in the worksheet. (When only part of a matrix is displayed, Mathcad adds a shaded border at the top and on the left side of the matrix, showing which rows and columns are displayed.) To see the values that are not displayed, you would click on the portion of the C matrix that is displayed, and scroll bars would appear. You could then scroll the matrix to see the remaining values. The scroll bars will appear only when a matrix has been selected (by clicking on the displayed portion of the matrix) and when the matrix is too large to display in its entirety.

Note: If you want to write the values in array C to a text file, you assign the array to the WRITEPRN( ) function:

   WRITEPRN("A:MyData.txt") := C

The values in C will be assigned to file MyData.txt on drive A:.

You can also use a *read file component* to read values from a data file into a Mathcad array. A read file component can be used to read ASCII text files, but it can also read spreadsheet and database files. A *component* is different than a simple function because it brings up a Wizard (a series of dialog boxes) that is used to set the parameters required to make the component function correctly. For a read file component, required parameters include the file location and file type. Optionally, you can instruct the component to read only a portion of the file.

To use a read file component, move the edit cursor to the location where the component should be placed, then insert the component using the menu commands Insert/Data/File Input....[3]

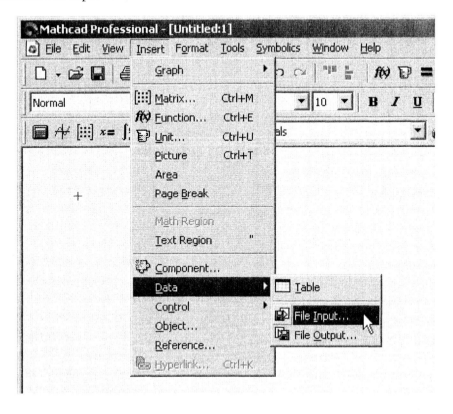

---

[3]In the latest version of Mathcad, the Insert menu has been rearranged slightly to make importing data files quicker. For versions of Mathcad prior to Mathcad 11 (2003), use Insert/Component.../File Read or Write to insert a read file component.

This starts the read file component Wizard so that you can specify the parameters necessary to read the file correctly.

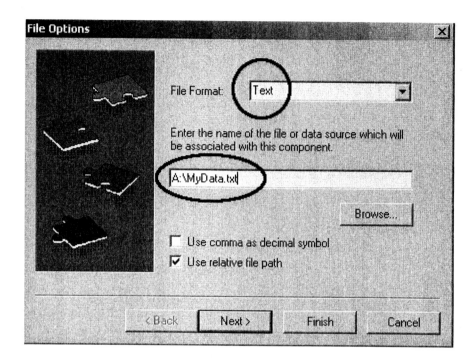

For this example, we again read the text file named `MyData.txt` on drive `A:`. Once the file type and location have been specified, click the Next button to continue to the next dialog box, where you can set the data range.

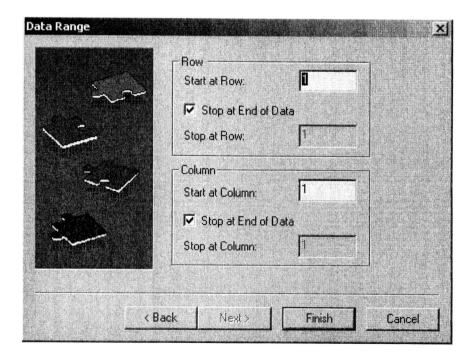

By simply accepting the default values on this dialog, the read file component will read all of the values in the file. Click the Finish button to complete the parameter specification and insert the read file component on the worksheet.

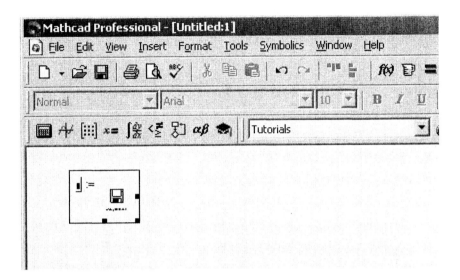

The icon for a read file component looks like a floppy disk. The placeholder on the left side of the assignment operator (:=) will hold the variable name of the array that is to receive the data. If we call the array D, and then click outside the region holding the read file component, Mathcad will read the values in the file and assign them to matrix D.

$$D := \quad \text{A:\textbackslash MyData.txt}$$

	0	1
0	0	50
1	10	48.2
2	20	46.5
3	30	44.8
4	40	43.2
5	50	41.6
6	60	40.1
7	70	38.7
8	80	37.3
9	90	35.9

$D =$

Again, only the first ten rows of matrix D are shown here. To see the remaining rows, click on any displayed portion of the matrix and scroll down.

Method 3: Using an Input Table to Create a Matrix

Mathcad has another *component* called an *input table* that provides a spreadsheet-like interface for entering arrays. This can be a convenient way to enter arrays by hand, but you can also automatically import values into an input table from various types of files (e.g., text files or spreadsheet files). Thus, the input table component is a convenient and flexible way of filling arrays with data from a variety of sources.

Note: The read file component is similar to the input table component because both can read a variety of file formats. However, there is a major distinction between the two components:

- When you import values into an input table, the file is read once and the values are copied into the Mathcad worksheet. Changes to values in the file are not reflected in the worksheet.
- A read file component reads the file each time the Mathcad worksheet is calculated (using Tools/Calculate/Calculate Worksheet), so changes to the values in the file *are* reflected in the worksheet.

To use an input table component to create and fill a matrix, do the following:[4]

1. Click on the worksheet to indicate where the input table should be located. Then insert the input table component by selecting Insert/Data/Table:

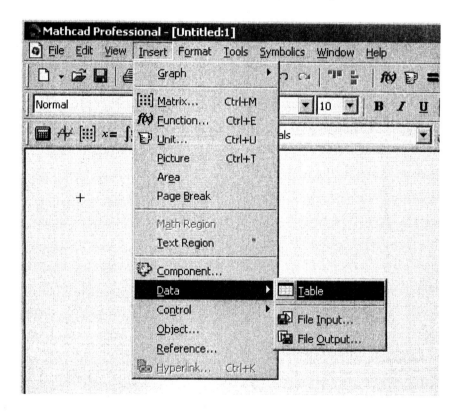

2. Enter the name of the array to be created in the placeholder at the top-left corner of the input table. In this example, the array will be called A:

---

[4] For versions of Mathcad prior to Mathcad 11 (2003), use Insert/Component ... then select Input Table from the Component Wizard dialog.

A :=

	0	1
0	0	
1		

Now you can simply start typing values into the table, or you can follow the remaining steps to import values from a file:

3. Click anywhere on the input table to select it.
4. Right-click on the table and select Import from the pop-up menu:

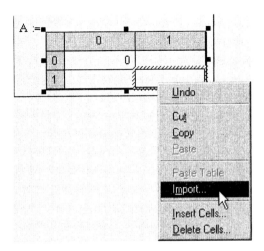

5. A File Options dialog will be displayed. Select the type of file (e.g., Excel file) to be imported, and then browse for the file.

Once the input table,

A :=

	0	1
0	1	1
1	2	2.46
2	3	4.17
3	4	6.06
4	5	8.1
5	6	10.27
6	7	12.55
7	8	14.93
8	8	17.4
9	10	19.95

has been filled, you can use array A for other calculations. Here, the filled array has simply been displayed as a matrix:

$$A = \begin{pmatrix} 1 & 1 \\ 2 & 2.462 \\ 3 & 4.171 \\ 4 & 6.063 \\ 5 & 8.103 \\ 6 & 10.271 \\ 7 & 12.550 \\ 8 & 14.929 \\ 9 & 17.399 \\ 10 & 19.953 \end{pmatrix}$$

Units on Matrix Elements

If all of the elements of an array have the same units, you can multiply the entire array by the units. If not, each element can have its own units. The two time vector definitions

$$\text{time} := \begin{bmatrix} 0 \\ 10 \\ 20 \\ 30 \end{bmatrix} \cdot \min$$

and

$$\text{time} := \begin{bmatrix} 0 \cdot \min \\ 10 \cdot \min \\ 20 \cdot \min \\ 0.5 \cdot \text{hr} \end{bmatrix}$$

are functionally equivalent.

Method 4a: Computing Array Element Values by Using Range Variables

Mathcad provides an unusual type of variable that takes on a whole series, or *range*, of values. This type of variable is called a *range variable* and can be used with arrays to identify particular elements of an array. For example, the first of the preceding time vectors could have been created using a range variable. The range variable needs to take the values 0, 1, 2, and 3, as it will be used to indicate each element of the time vector in turn. The range variable can have any name, but i and j are often employed, since range variables are commonly used as index variables. The range variable is defined as [i][:][0][;][3], or

    i := 0..3

Note that you indicate the range by using [;] (semicolon), but Mathcad displays a series of dots (called an *ellipsis*). The range variable we've defined has four values, and anytime you use this variable in an equation, the equation will be evaluated four times, once with i = 0, once with i = 1, and so on until the end of the range. Range variables can be very handy, but they do take a little getting used to.

Once the range variable has been defined, the elements of the time vector can be calculated with the use of the range variable, like this:

$$time_i := (i \cdot 10) \cdot min$$

$$time = \begin{bmatrix} 0 \\ 10 \\ 20 \\ 30 \end{bmatrix} \cdot min$$

The range variable has been used as an index subscript (left-square-bracket subscript) on the left side of the definition, to indicate which element of the time vector is being computed, and is also used on the right side as part of the calculation itself. As the value of the range variable changes from 0 to 3, the computed element values change from 0 to 30 minutes.

PRACTICE!

What will the matrix M look like when defined by the following expressions?

a. $i := 0..4$
   $M_i := 2 \cdot i$

b. $i := 0..4$
   $j := 0..3$
   $M_{i,j} := (3 + 2 \cdot i - j) \cdot ft$

An Example of a Two-Dimensional Array

In the following equations, range variables r and c are defined to indicate the row and column of the element of the matrix S being computed, and the value of the element depends on the values of r and c:

$$r := 0..4 \quad c := 0..2$$
$$S_{r,c} := r^2 + c^2$$

$$S = \begin{bmatrix} 0 & 1 & 4 \\ 1 & 2 & 5 \\ 4 & 5 & 8 \\ 9 & 10 & 13 \\ 16 & 17 & 20 \end{bmatrix}$$

Method 4b: Computing Array Element Values Using the Matrix( ) Function

Mathcad's matrix( ) function allows you to fill an array with computed values without explicitly declaring range variables. In effect, Mathcad declares the range variables on the basis of the information you send to the matrix( ) function. The matrix(r, c, f) function creates a matrix with r rows and c columns in which the value of the i,jth element is computed using the function f(). This function must be a function of two variables and must be declared before calling the matrix( ) function. The function f(i,j) is evaluated repeatedly to fill the matrix, with i ranging from 0 to r−1 and j ranging from 0 to c−1.

The S matrix could be created using the matrix( ) function as follows:

$$myFunc(r,c) := r^2 + c^2$$
$$S := matrix(5, 3, myFunc)$$

$$S = \begin{bmatrix} 0 & 1 & 4 \\ 1 & 2 & 5 \\ 4 & 5 & 8 \\ 9 & 10 & 13 \\ 16 & 17 & 20 \end{bmatrix}$$

PRACTICE!

What would the matrix returned by matrix(4, 4, f) look like, if f(r, c) were defined as follows?

a. f(r, c) := 2 + r + c
b. f(r, c) := 0.5·r + c²

The following matrix was created with the use of the matrix() function:

M := matrix(3, 5, f)

$$M = \begin{bmatrix} 0 & 3 & 6 & 9 & 12 \\ 1 & 4 & 7 & 10 & 13 \\ 2 & 5 & 8 & 11 & 14 \end{bmatrix}$$

What did f(r, c) look like?

Method 5: Copying and Pasting Arrays from Other Windows Programs

A convenient way to move values from another program into Mathcad is to copy the values in the other program (e.g., a spreadsheet) and then paste them into an array definition in Mathcad. If the values already exist in another Windows program, this can save a lot of typing.

In the next example, two columns of values are displayed in an Excel spreadsheet as follows:

	A	B	C	D
				=A1^3
1	1	1		
2	2	8		
3	3	27		
4	4	64		
5	5	125		
6				

The values shown in column B are computed by raising the values in column A to the third power. (You do not need to convert the spreadsheet formulas to values before moving the data to Mathcad.)

To create a 5 × 2 array in Mathcad containing these values, select the 10 values in cells A1..B5, and copy them to the Windows clipboard (using the spreadsheet's menu commands Edit/Copy, for example). Then begin an array definition in Mathcad:

C := ▮

Click on the placeholder, and paste the contents of the clipboard into the placeholder by using Mathcad's Edit/Paste menu commands.

The values displayed in the spreadsheet are now a Mathcad array:

$$C := \begin{bmatrix} 1 & 1 \\ 2 & 8 \\ 3 & 27 \\ 4 & 64 \\ 5 & 125 \end{bmatrix}$$

Note: The formulas used in the spreadsheet did not come across to Mathcad; just the displayed values were pasted.

PROFESSIONAL SUCCESS

Let your software do the work.

Retyping data, equations, and results into a report is tedious and tends to introduce errors. The past few generations of software products have had the ability to exchange data, and the capabilities of the products are improving with every new version. Mathcad can share information with other mathematics packages, such as Excel and Matlab (data), and word processors such as Microsoft Word (values, equations, graphs).[1] If you find yourself frequently reentering the same information into multiple software packages you might want to see if there's a better way.

---
[1] Microsoft Word and Excel are products of the Microsoft Corporation, Redmond, WA. Matlab is produced by The Mathworks, Inc., Natick, MA.

Creating an Identity Matrix

An *identity matrix* is a square matrix filled with ones along the diagonal and zeroes everywhere else. Mathcad's `identity( )` function creates identity matrices. Since the number of rows is always equal to the number of columns in an identity matrix, you need to specify just one or the other, so the `identity( )` function takes only a single parameter indicating the number of rows (or columns). Here's an example of the use of the `identity()` function to create a 5 × 5 identity matrix called `ID`:

```
ID := identity(5)
```

$$ID = \begin{bmatrix} 1 & 0 & 0 & 0 & 0 \\ 0 & 1 & 0 & 0 & 0 \\ 0 & 0 & 1 & 0 & 0 \\ 0 & 0 & 0 & 1 & 0 \\ 0 & 0 & 0 & 0 & 1 \end{bmatrix}$$

PRACTICE!

Create a 5 × 5 identity matrix by using

```
ID := identity(5)
```

Then use the column operator to pull out the center column of matrix `ID` (column number 2). What would change when you used the column operator to extract the column number 2 if the identity matrix were created like this?

```
ORIGIN := 1
ID := identity(5)
```

4.3 MODIFYING AN ARRAY

Suppose you have just entered a large array from the keyboard and notice that you accidentally left out one row. You don't have to start all over again; you can insert a row or column into an existing array. You can also delete one or more rows or columns. In addition, Mathcad allows you to join arrays either side to side [augment( )] or one on top of the other [stack( )]. Finally, Mathcad lets you assign portions of an array to a new variable by using the submatrix( ) function.

Inserting a Row or Column into an Existing Array

Suppose that you have just entered a massive array by hand and discovered that you left out a row. Don't worry; all is not lost. Mathcad will allow you to insert one or more rows into an existing array. Mathcad inserts the row after the currently selected row, so click on the row immediately above where you want the new row to be inserted. For example, to insert a new row just after the row containing 3, 9, and 27, click on the 3, 9, or 27 to indicate where the new row should be placed:

$$G := \begin{bmatrix} 1 & 1 & 1 \\ 2 & 4 & 8 \\ 3 & 9 & 2|7 \\ 4 & 16 & 64 \end{bmatrix}$$

Then bring up the Insert Matrix dialog box by pressing [Ctrl-M]. Indicate the number of rows to be added (and zero columns!), and then press the Insert or OK button:

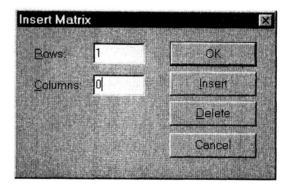

A row of placeholders is inserted into the matrix:

$$G := \begin{bmatrix} 1 & 1 & 1 \\ 2 & 4 & 18 \\ 3 & 9 & 27 \\ \blacksquare & \blacksquare & \blacksquare \\ 4 & 16 & 64 \end{bmatrix}$$

Click on the placeholders to enter values.

For example, to add two columns at the right side of the original G matrix, do the following:

1. Click on any element in the third column. Mathcad will insert the new columns to the right of the column you select:

$$G := \begin{bmatrix} 1 & 1 & 1 \\ 2 & 4 & 8 \\ 3 & 9 & 27 \\ 4 & 16 & 64 \end{bmatrix}$$

2. Bring up the Insert Matrix dialog box by pressing [Ctrl-M]. Indicate the number of columns to be added (and zero rows), and then press the Insert button:

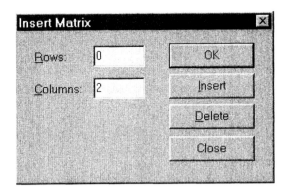

The result will be

$$G := \begin{bmatrix} 1 & 1 & 1 & \blacksquare & \blacksquare \\ 2 & 4 & 8 & \blacksquare & \blacksquare \\ 3 & 9 & 27 & \blacksquare & \blacksquare \\ 4 & 16 & 64 & \blacksquare & \blacksquare \end{bmatrix}$$

To add rows at the top of the array or columns at the left, select the entire array, rather than a single element, before inserting rows or columns:

$$G := \begin{Vmatrix} 1 & 1 & 1 \\ 2 & 4 & 8 \\ 3 & 9 & 27 \\ 4 & 16 & 64 \end{Vmatrix}$$

The matrix should now look like this:

$$G := \begin{bmatrix} \blacksquare & \blacksquare & \blacksquare & \blacksquare \\ \blacksquare & 1 & 1 & 1 \\ \blacksquare & 2 & 4 & 8 \\ \blacksquare & 3 & 9 & 27 \\ \blacksquare & 4 & 16 & 64 \end{bmatrix}$$

The preceding example illustrates that Mathcad will allow the addition of both rows and columns in a single operation.

Deleting Rows or Columns

The procedure for deleting rows and columns is similar to that used to insert columns. To delete one or more rows, select an element in the first row or column to be deleted. If you want to delete three rows, the rows will include the row containing the selected element and the two rows immediately below it. Similarly, to delete two columns, select an element in the leftmost column to be removed. The deleted columns will include the column containing the selected element and the column to the right of it. For example, to delete the middle two rows of the original G array, do the following:

1. Select an element in the second row of the array:

$$G := \begin{bmatrix} 1 & 1 & 1 \\ 2| & 4 & 8 \\ 3 & 9 & 27 \\ 4 & 16 & 64 \end{bmatrix}$$

2. Bring up the Insert Matrix dialog box. Set the number of rows to be deleted to 2 and the number of columns to 0. Click the Delete button on the dialog box:

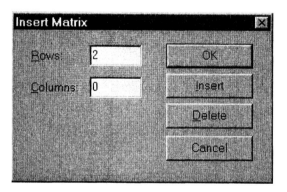

The middle two rows of the G matrix have been deleted:

$$G := \begin{bmatrix} 1 & 1 & 1 \\ 4 & 16 & 64 \end{bmatrix}$$

Selecting a Portion of an Array

There are times when you need just a portion of an array. Mathcad provides a *column operator* [Ctrl-6] that allows you to select a single column and a `submatrix( )` function that allows you to select an arbitrary subsection of an array.

Selecting a Single Column Mathcad's *column operator* allows you to take a single column (vector) of a multicolumn array and assign that column to a new variable. The column operator is available on the Matrix Toolbar or by pressing [Ctrl-6]. To see how the column operator is used, consider the time and concentration data that were read from the `MyData.txt` file:

$$C := \text{READPRN}(\text{"A:MyData.txt"})$$

$$C = \begin{array}{c|cc} & 0 & 1 \\ \hline 0 & 0 & 50 \\ 1 & 10 & 48.2 \\ 2 & 20 & 46.5 \\ 3 & 30 & 44.8 \\ 4 & 40 & 43.2 \\ 5 & 50 & 41.6 \\ 6 & 60 & 40.1 \\ 7 & 70 & 38.7 \\ 8 & 80 & 37.3 \\ 9 & 90 & 35.9 \end{array}$$

The first column actually contains time information in minutes, and the second column contains concentration data in units of mg/L. Because the units on the columns are different, we can't just multiply the entire array by a set of units. But if we separate the time and concentration vectors, then we can put units on each vector.

We can use the column operator to extract the left column (column 0) and assign the values to a new variable, called `time`. We can build in units at the same time:

$$\text{time} := C^{<0>} \cdot \text{min}$$

$$\text{time} = \begin{array}{c|c} & 0 \\ \hline 0 & 0 \\ 1 & 10 \\ 2 & 20 \\ 3 & 30 \\ 4 & 40 \\ 5 & 50 \\ 6 & 60 \\ 7 & 70 \\ 8 & 80 \\ 9 & 90 \end{array} \cdot \text{min}$$

Similarly, we can pull out the concentration data and assign it to a new variable, `conc`:

$$\text{conc} := C^{<1>} \cdot \frac{\text{mg}}{\text{L}}$$

$$\text{conc} = \begin{array}{|c|c|} \hline & 0 \\ \hline 0 & 50 \\ 1 & 48.2 \\ 2 & 46.5 \\ 3 & 44.8 \\ 4 & 43.2 \\ 5 & 41.6 \\ 6 & 40.1 \\ 7 & 38.7 \\ 8 & 37.3 \\ 9 & 35.9 \\ \hline \end{array} \cdot \frac{\text{mg}}{\text{L}}$$

This is an easy way to get units on values read in from text files. The column operator is also useful in getting vectors ready for plotting, since Mathcad allows you to plot one vector against another simply by entering the vector names in the placeholders on the x- and y-axes of an $x - y$ plot:

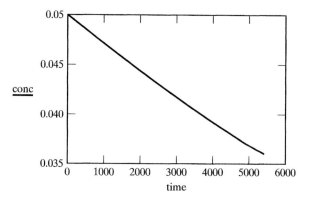

The values on the axes may surprise you, but remember that Mathcad always displays graphs in base units, so the concentrations are in kg/m^3, and the times are in seconds.

Note: You can "trick" Mathcad into displaying values in the other units by dividing the vector names (on the axes) by the desired units. The resulting graph appears to display the correct values (but the axis labels can be quite misleading):

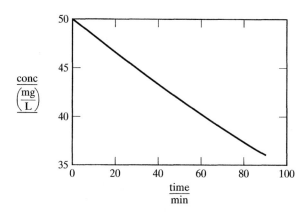

Selecting a Single Row Mathcad does not provide a mechanism for selecting a single row, but you can get the job done by turning the array sideways (i.e., transposing the array) and selecting a column by means of the column operator. Going back to the G matrix that we used before, suppose we want to select the third row and assign it to a new variable, R3. We have

$$G := \begin{bmatrix} 1 & 1 & 1 \\ 2 & 4 & 8 \\ 3 & 9 & 27 \\ 4 & 16 & 64 \end{bmatrix}$$

We need to turn the G array around by using the transpose operator from the Matrix Toolbar:

$$G_{tr} := G^T$$

$$G_{tr} = \begin{bmatrix} 1 & 2 & 3 & 4 \\ 1 & 4 & 9 & 16 \\ 1 & 8 & 27 & 64 \end{bmatrix}$$

Then we can select what is now the third column (Mathcad's column number 2, since it starts counting at zero) by using the column operator:

$$R3_{tr} := G_{tr}^{<2>}$$

$$R3_{tr} = \begin{bmatrix} 3 \\ 9 \\ 27 \end{bmatrix}$$

Finally, we turn the result back into a row vector by transposing the $R3_{tr}$ vector:

$$R3 := R3_{tr}^T$$
$$R3 = [3 \quad 9 \quad 27]$$

This multistep operation can be carried out in a single command, thereby eliminating some variables:

$$R3 := [(G^T)^{<2>}]^T$$
$$R3 = [3 \quad 9 \quad 27]$$

If you need to extract a row frequently, you might want to create a function that operates like a "row operator":

$$\text{row\_operator}(A, r) := (A^T)^{<r>T}$$

This function can then be used to extract the third row of the G matrix:

$$R3 := \text{row\_operator}(G, 2)$$
$$R3 = [3 \quad 9 \quad 27]$$

Choosing a Subset of an Array An alternative to all of the transposing that was done in the preceding example is Mathcad's submatrix(A, r_{start}, r_{stop}, c_{start}, c_{stop}) function. This function takes a number of parameters and allows you to specify exactly which part of a matrix to extract. The first parameter, A, is the name of the array from which the submatrix is to be taken. The starting and stopping row and column numbers must be integers. Remember that, by default, Mathcad calls the top-left element, $A_{0,0}$. If you wanted to start with that element, then r_{start} and c_{start} would both be zero, not one.

To choose the third row of the G matrix, the submatrix( ) function could be used with the following arguments:

```
R3 := submatrix(G, 2, 2, 0, 2)
```

These arguments tell the submatrix( ) function to start with the G array and pull out rows 2 through 2 (using Mathcad's way of counting rows) and columns 0 through 2:

```
R3 := submatrix(G,2,2,0,2)
R3 = [3  9  27]
```

The submatrix( ) function can also return portions of an array that are not complete rows or columns. For example, the function could be used to extract the four elements in the top-left corner of the G matrix:

```
TL4 := submatrix(G,0,1,0,1)
```

$$TL4 = \begin{bmatrix} 1 & 1 \\ 2 & 4 \end{bmatrix}$$

Combining Two Arrays

Mathcad provides two functions for combining arrays: augment( ) and stack( ). The stack( ) function stacks matrices, and the augment( ) function connects them side by side. Here are a couple of simple examples:

$$A := \begin{bmatrix} 1 & 2 & 3 \\ 2 & 3 & 4 \\ 3 & 4 & 5 \end{bmatrix} \qquad B := \begin{bmatrix} 7 & 8 & 9 \\ 8 & 9 & 10 \\ 9 & 10 & 11 \end{bmatrix}$$

```
Aug := augment(A,B)
```

$$Aug := \begin{bmatrix} 1 & 2 & 3 & 7 & 8 & 9 \\ 2 & 3 & 4 & 8 & 9 & 10 \\ 3 & 4 & 5 & 9 & 10 & 11 \end{bmatrix}$$

```
Stk := stack(A,B)
```

$$Stk = \begin{bmatrix} 1 & 2 & 3 \\ 2 & 3 & 4 \\ 3 & 4 & 5 \\ 7 & 8 & 9 \\ 8 & 9 & 10 \\ 9 & 10 & 11 \end{bmatrix}$$

PRACTICE!

In what order would you use the `stack()` and `augment()` functions to create the following matrix from matrices A and B in the previous section?

$$T = \begin{bmatrix} 1 & 2 & 3 & 1 & 2 & 3 \\ 2 & 3 & 4 & 2 & 3 & 4 \\ 3 & 4 & 5 & 3 & 4 & 5 \\ 7 & 8 & 9 & 1 & 2 & 3 \\ 8 & 9 & 10 & 2 & 3 & 4 \\ 9 & 10 & 11 & 3 & 4 & 5 \end{bmatrix}$$

How would the `submatrix()` function be used to extract matrix

$$T_{sub} = \begin{bmatrix} 5 & 3 & 4 \\ 9 & 1 & 2 \\ 10 & 2 & 3 \\ 11 & 3 & 4 \end{bmatrix}$$

from the T matrix?

4.4 ARRAY OPERATIONS

Array operations such as addition and multiplication require that certain conditions be met and that certain procedures be followed. For each of the standard array operations that follow, the requirements and the procedures are listed.

Array Addition and Subtraction

Requirement: The arrays to be added or subtracted must be the same size.

Procedure: Element-by-element addition.

Array addition and subtraction in Mathcad are handled just as are scalar addition and subtraction:

$$A := \begin{bmatrix} 1 & 2 & 3 \\ 2 & 3 & 4 \\ 3 & 4 & 5 \end{bmatrix} \quad B := \begin{bmatrix} 7 & 8 & 9 \\ 8 & 9 & 10 \\ 9 & 10 & 11 \end{bmatrix}$$

Sum := A+B

$$Sum = \begin{bmatrix} 8 & 10 & 12 \\ 10 & 12 & 14 \\ 12 & 14 & 16 \end{bmatrix}$$

Dif := Sum−B

$$Dif = \begin{bmatrix} 1 & 2 & 3 \\ 2 & 3 & 4 \\ 3 & 4 & 5 \end{bmatrix}$$

Note: Array subtraction is not a standard array operation; you usually have to multiply by −1 and then add. Array subtraction, however, is allowed in Mathcad.

Matrix Multiplication

Requirement: The inside dimensions of the arrays to be multiplied must be equal. The outside dimensions determine the size of the product matrix. Thus, we might have

$D_{2 \times 3} \times E_{3 \times 2}$ inner dimensions are equal (3)
       product dimensions will be 2×2

Procedure: Working across the columns of the first array and down the rows of the second array, multiply elements and add the results. Mathematically, matrix multiplication is summarized as

```
Prod₀,₀ = [(1×10)+(2+12)+(3×14)]=76
Prod₀,₁ = [(1×11)+(2×13)+(3×15)]=82
Prod₁,₀ = [(4×10)+(5×12)+(6×14)]=184
Prod₁,₁ = [(4×11)+(5×13)+(6×15)]=199
```

$$D := \begin{bmatrix} 1 & 2 & 3 \\ 4 & 5 & 6 \end{bmatrix} \qquad E := \begin{bmatrix} 10 & 11 \\ 12 & 13 \\ 14 & 15 \end{bmatrix}$$

Prod := D·E

$$\text{Prod} = \begin{bmatrix} 76 & 82 \\ 84 & 199 \end{bmatrix}$$

PRACTICE!

> The order in which the matrices are multiplied makes a difference. What is the result of the following matrix multiplication?
>
> Prod₂ := E·D

Element-by-Element Multiplication

Sometimes you don't want true matrix multiplication. Instead, you want each element of the first matrix multiplied by the corresponding element of the second matrix. Element-by-element multiplication is available in Mathcad by means of the *vectorize* operator on the Matrix Toolbar.

Requirement: The arrays must be the same size for element-by-element multiplication.

Procedure: Multiply each individual element of the first matrix by the corresponding element of the second matrix:

$$A := \begin{bmatrix} 1 & 2 & 3 \\ 2 & 3 & 4 \\ 3 & 4 & 5 \end{bmatrix} \quad B := \begin{bmatrix} 7 & 8 & 9 \\ 8 & 9 & 10 \\ 9 & 10 & 11 \end{bmatrix}$$

$$\text{ElemMult} = \overrightarrow{(A \cdot B)}$$

$$\text{ElemMult} = \begin{bmatrix} 7 & 16 & 27 \\ 16 & 27 & 40 \\ 27 & 40 & 55 \end{bmatrix}$$

PRACTICE!

Compare the results of the following matrix multiplications:

$A \cdot B$

$\overrightarrow{(A \cdot B)}$

Transposition

Requirement: Any array can be transposed.

Procedure: Interchange row and column elements, $\text{Trans}_{j,i} = C_{i,j}$:

$$C := \begin{bmatrix} 1 & 1 \\ 2 & 8 \\ 3 & 27 \\ 4 & 64 \\ 5 & 125 \end{bmatrix}$$

$\text{Trans} := C^T$

$$\text{Trans} = \begin{bmatrix} 1 & 2 & 3 & 4 & 5 \\ 1 & 8 & 27 & 64 & 125 \end{bmatrix}$$

Inversion

When you invert a scalar, you divide 1 by the scalar, and the result is a value that, when multiplied by the original value, yields 1 as a product. Similarly, when you multiply an inverted matrix by the original matrix, you obtain an identity matrix as a result.

Requirement: Only square matrices can be inverted, and the matrix must be nonsingular.

Procedure: The procedure for inverting a matrix is quite involved. You first augment the matrix with an identity matrix and then use row operations by multiplying by scale factors and then adding and subtracting rows to convert the original matrix into an identity matrix. The same row operations will transform the original identity matrix into the inverse matrix. (For more details, see any text on matrix mathematics.) Suffice it to say that, for large matrices, the inversion process requires many calculations, and there can be significant round-off error associated with the process. How big is a "large" matrix? That depends on how accurate you want your solutions to be, but when matrices get above about 20×20, you should start trying to find solution methods that avoid inversions.

Inverting a matrix in Mathcad is easy; you simply raise the matrix to the -1 power:

$$F := \begin{bmatrix} 2 & 3 & 5 \\ 7 & 2 & 4 \\ 8 & 11 & 6 \end{bmatrix}$$

$$F^{-1} = \begin{bmatrix} -0.1517 & 0.1754 & 0.0095 \\ -0.0474 & -0.1327 & 0.128 \\ 0.2891 & 0.0095 & -0.0806 \end{bmatrix}$$

If Mathcad cannot invert the matrix, it will tell you that the matrix is singular.

A common use of matrix inversion is in the solution of systems of linear equations. A system of linear equations can be written in matrix form, consisting of a coefficient matrix, an unknown vector, and a right-hand-side vector:

$$[C][x] = [r]$$

One common solution method requires that the coefficient matrix be inverted and multiplied by the right-hand-side vector to calculate the values of the unknowns:

$$[x] = [C]^{-1}[r]$$

If the coefficient matrix is singular—that is, if it cannot be inverted—then there is no unique solution to the original set of equations. If there is a solution to the equations, you should be able to invert the coefficient matrix.

APPLICATIONS: REMOVAL OF CO_2 FROM A GAS STREAM

Solutions of various ethanolamines [(e.g., monoethanolamine (MEA) or diethanolamine (DEA)] or potassium carbonate in water are commonly used to remove CO_2 from gas streams. The solution contacts the gas stream in a tall tower. The liquid flows down while the gas flows up, and the tower is designed to provide good contact between the liquid and the gas.

One hundred moles of a gas stream containing 10% CO_2 and 90% other combustion products (OCP) is fed at the bottom of the tower, and a solution (SOLN) containing 2% CO_2 and 98% MEA is fed at the top (s_1).

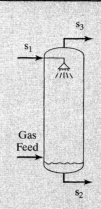

The exiting streams are analyzed and found to contain the following products:

liquid out (s_2): 12% CO_2 1% OCP 87% SOLN
gas out (s_3): 1% CO_2 99% OCP

How many moles are in each of the unknown streams s_1, s_2, and s_3?

SOLUTION

This problem can be solved by writing material balances. The material balance for CO_2, for example, simply states that the CO_2 going into the process (10% of the 100 moles of gas in and 2% of the liquid in) must come out again (1% of the gas out and 12% of the liquid out). In mathematical terms, we have

CO_2 Balance: $10 + 0.02 \cdot s_1$
$= 0.12 \cdot s_2 + 0.01 \cdot s_3$

The idea that the CO_2 entering the tower must leave again is true if the process operates at steady state (CO_2 is not accumulating in the tower) and if there is no chemical reaction involving CO_2 taking place in the tower.

Similar balances can be written for OCP and SOLN:

OCP Balance: $90 = 0.01 \cdot s_2 + 0.99 \cdot s_3$
SOLN Balance: $0.98 \cdot s_1 1 = 0.87 \cdot s_2$

To solve for the moles in s_1, s_2, and s_3, we first collect all the terms involving an 's' on one side and all the constants on the other:

$$0.02 \cdot s_1 - 0.12 \cdot s_2 - 0.01 \cdot s_3 = -90$$
$$-0.01 \cdot s_2 - 0.99 \cdot s_3 = -90$$
$$0.98 \cdot s_1 - 0.87 \cdot s_2 = 0$$

This set of equations can be written in matrix form as a coefficient matrix C and a right-hand-side vector r. Using Mathcad, we define the matrices as follows:

$$C := \begin{bmatrix} 0.02 & -0.12 & -0.01 \\ 0 & -0.01 & -0.99 \\ 0.98 & -0.87 & 0 \end{bmatrix} \quad r := \begin{bmatrix} -10 \\ -90 \\ 0 \end{bmatrix}$$

We solve for the s vector (the numbers of moles in the unknown streams) by inverting the C matrix and multiplying the result with the r vector:

$$C_{inv} := C^{-1} \quad C_{inv}$$
$$= \begin{bmatrix} -8.691 & 0.088 & 1.198 \\ -9.79 & 0.099 & 0.2 \\ 0.099 & -1.011 & -2.018 \cdot 10^{-3} \end{bmatrix}$$

$$s := C_{inv} \cdot r \quad s = \begin{bmatrix} 79 \\ 89 \\ 90 \end{bmatrix}$$

Stream s_1 contains 79 moles (77.4 moles SOLN, 1.6 moles CO_2), stream s_2 contains 89 moles, and stream s_3 contains 90 moles.

Determinant

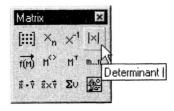

The *determinant* is a scalar value that can be computed from a square matrix. The process for computing a determinant is fairly straightforward, but tedious for matrices larger than 3 × 3. Fortunately, Mathcad will compute determinants automatically. The determinant operator is found on the Matrix Toolbar.

For a 1 × 1 matrix (a single value), the determinant of the matrix is simply the value of the matrix. Vertical bars are used to indicate the determinant, as, for example, in

$$|4| = 4$$

For a 2 × 2 matrix, the determinant is computed by multiplying diagonal elements and subtracting the results:

$$A := \begin{bmatrix} A_{00} & A_{01} \\ A_{10} & A_{11} \end{bmatrix}$$

$$D = |A| = A_{00} \cdot A_{11} - A_{10} \cdot A_{01}$$

For larger matrices, the determinants are found by breaking down the matrix into smaller units called *cofactors*, calculating the determinants for each cofactor, and summing the results. The general equation for computing a determinant is

$$D = A_{i0} + C_{i0} + A_{i1}C_{i1} + \cdots + A_{i(n-1)}C_{i(n-1)} \quad i = 0 \ldots (n-1)$$

where the C's are the cofactors. For a 3 × 3 matrix, the cofactors are computed as follows:

- The cofactor for element A_{00} is computed from the shaded terms in the following matrix and is $C_{00} = +(A_{11} \cdot A_{22} - A_{21} \cdot A_{12})$ (*note the plus sign before this cofactor*):

  $$\begin{matrix} \boxed{A_{00}} & A_{01} & A_{02} \\ A_{10} & A_{11} & A_{12} \\ A_{20} & A_{21} & A_{22} \end{matrix}$$

- For element A_{10}, the cofactor is $C_{10} = -(A_{01} \cdot A_{22} - A_{21} \cdot A_{02})$ (*note the minus sign before this cofactor*):

  $$\begin{matrix} A_{00} & A_{01} & A_{02} \\ \boxed{A_{10}} & A_{11} & A_{12} \\ A_{20} & A_{21} & A_{22} \end{matrix}$$

- For element A_{20}, the cofactor is $C_{20} = +(A_{01} \cdot A_{12} - A_{11} \cdot A_{02})$ (*note the plus sign before this cofactor*):

  $$\begin{matrix} A_{00} & A_{01} & A_{02} \\ A_{10} & A_{11} & A_{12} \\ \boxed{A_{20}} & A_{21} & A_{22} \end{matrix}$$

- The sign in front of the cofactor equation alternates between plus and minus, depending on the position of the cofactor in the matrix, with $C_{0,0}$ (the top-left element) always taking the plus sign:

  $$\begin{matrix} \boxed{+} & - & + \\ - & + & - \\ + & - & + \end{matrix}$$

With numbers, the process looks like this:

$$F := \begin{bmatrix} 2 & 3 & 4 \\ 7 & 2 & 5 \\ 8 & 11 & 6 \end{bmatrix}$$

$C_{0,0} := 2 \cdot 6 - 11 \cdot 4$ $\qquad\qquad C_{0,0} = -32$

$C_{1,0} := -(3 \cdot 6 - 11 \cdot 5)$ $\qquad\qquad C_{1,0} = 37$

$C_{2,0} := 3 \cdot 4 - 2 \cdot 5$ $\qquad\qquad C_{2,0} = 2$

$D := F_{0,0} \cdot C_{0,0} + F_{1,0} \cdot C_{1,0} + F_{2,0} \cdot C_{2,0}$ $\qquad D = 211$

Or, using the determinant operator from the Matrix Toolbox, we obtain

$$|F| = 211$$

Note: In the preceding example, the determinant was computed using cofactors for elements in the left column. Actually, you can use cofactors for the elements in any column or row to compute the determinant.

How Is the Determinant Used?

The determinant shows up in a number of engineering calculations, but one of the most straightforward applications is using a determinant to find out whether a system of simultaneous linear equations has a unique solution. If the determinant of the coefficient matrix is nonzero, then the system has a unique solution. For example, we could have checked whether the coefficient matrix in the last Application example could be solved by calculating its determinant:

$$C := \begin{bmatrix} 0.02 & -0.12 & -0.01 \\ 0 & -0.01 & -0.99 \\ 0.98 & -0.87 & 0 \end{bmatrix} \quad |C| = 0.099$$

Because the determinant is nonzero, there should be a solution, and there is.

Note: The calculation of determinants suffers from the same problem encountered with matrix inversion: For a large matrix, the number of calculations required leads to round-off errors on digital computers.

PRACTICE!

Each of the two systems of equations that follow has no unique solution. Verify this statement by calculating the determinant of the coefficient matrix.

a. The second and third equations are identical:

$$2 \cdot x_1 + 3 \cdot x_2 + 1 \cdot x_3 = 12$$
$$1 \cdot x_1 + 4 \cdot x_2 + 7 \cdot x_3 = 16$$
$$1 \cdot x_1 + 4 \cdot x_2 + 7 \cdot x_3 = 16$$

b. The third equation is the sum of the first and second equations:

$$2 \cdot x_1 + 3 \cdot x_2 + 1 \cdot x_3 = 12$$
$$1 \cdot x_1 + 4 \cdot x_2 + 7 \cdot x_3 = 16$$
$$3 \cdot x_1 + 7 \cdot x_2 + 8 \cdot x_3 = 28$$

PRACTICE!

The two systems of equations that follow might have solutions. Use the determinant to find out for sure whether they do.

a. $2 \cdot x_1 + 3 \cdot x_2 + 1 \cdot x_3 = 12$
$1 \cdot x_1 + 4 \cdot x_2 + 7 \cdot x_3 = 16$
$4 \cdot x_1 + 1 \cdot x_2 + 3 \cdot x_3 = 9$

b. $2 \cdot x_1 + 3 \cdot x_2 + 1 \cdot x_3 = 12$
$1 \cdot x_1 + 4 \cdot x_2 + 7 \cdot x_3 = 16$
$7 \cdot x_1 + 18 \cdot x_2 + 23 \cdot x_3 = 48$

4.5 ARRAY FUNCTIONS

Several functions are sometimes useful when one is working with arrays. The `min(A)` and `max(A)` functions find the minimum and maximum values in the array specified as the function's parameter:

$$C := \begin{bmatrix} 1 & 1 \\ 2 & 8 \\ 3 & 27 \\ 4 & 64 \\ 5 & 125 \end{bmatrix} \quad \begin{array}{l} \max(C) = 125 \\ \min(C) = 1 \end{array}$$

$$a := \begin{bmatrix} 3 \\ 2 \\ 7 \\ 4 \end{bmatrix} \quad \begin{array}{l} \max(a) = 7 \\ \min(a) = 2 \end{array}$$

Four additional functions return information on the size of an array or a vector. The `cols(A)` function returns the number of columns in array A, while the `rows(A)` function returns the number of rows in A. For vectors, there is a `length(v)` function that returns the number of elements in the vector and a `last(v)` function that returns the index of the last element of the vector. For example, we might have

$$C := \begin{bmatrix} 1 & 1 \\ 2 & 8 \\ 3 & 27 \\ 4 & 64 \\ 5 & 125 \end{bmatrix} \quad \begin{array}{l} \text{rows}(C) = 5 \\ \text{cols}(C) = 2 \end{array}$$

$$a := \begin{bmatrix} 3 \\ 2 \\ 7 \\ 4 \end{bmatrix} \quad \begin{array}{l} \text{rows}(a) = 4 \\ \text{cols}(a) = 1 \\ \text{length}(a) = 4 \\ \text{last}(a) = 3 \quad a_3 = 4 \end{array}$$

Note that the `rows(A)` and `cols(A)` functions work with either matrices or vectors, but the `length(v)` and `last(v)` functions operate only on vectors. The value returned by `last(v)` will be one less than that returned by `length(v)` as long as the matrix ORIGIN is zero.

Sorting

There are three functions for sorting vectors and arrays in Mathcad. The `sort(v)` function arranges the elements of a (v) vector in increasing order. You can combine the `sort(v)` function with the `reverse(v)` function to get the elements of a vector arranged in decreasing order:

$$a_1 := \text{sort}(a)$$

$$a_1 := \begin{bmatrix} 2 \\ 3 \\ 4 \\ 7 \end{bmatrix}$$

$$a_2 := \text{reverse}(a_1)$$

$$a_2 := \begin{bmatrix} 7 \\ 4 \\ 3 \\ 2 \end{bmatrix}$$

Note that the reverse(v) function is not a sorting function: It doesn't perform a sort, but just reverses the order of the elements in a vector.

There are two sorting functions for arrays: csort(A,n) and rsort(A,n). The csort(A,n) function arranges the rows of array A such that the elements in column n will be in increasing order. To sort array H on the left column, use the command csort(H,0), as shown here:

$$H := \begin{bmatrix} 7 & 9 & 2 \\ 4 & 8 & 1 \\ 8 & 2 & 0 \\ 3 & 7 & 4 \end{bmatrix}$$

$$H_1 := \text{csort}(H, 0) \qquad H_1 = \begin{bmatrix} 3 & 7 & 4 \\ 4 & 8 & 1 \\ 7 & 9 & 2 \\ 8 & 2 & 0 \end{bmatrix}$$

To sort array H on the rightmost column, use the command csort (H,2):

$$H_2 := \text{csort}(H, 2) \qquad H_2 = \begin{bmatrix} 8 & 2 & 0 \\ 4 & 8 & 1 \\ 7 & 9 & 2 \\ 3 & 7 & 4 \end{bmatrix}$$

To rearrange the columns so that the elements in the top row are in increasing order, use the rsort(A,0) command:

$$H_3 := \text{rsort}(H, 0) \qquad H_3 = \begin{bmatrix} 2 & 7 & 9 \\ 1 & 4 & 8 \\ 0 & 8 & 2 \\ 4 & 3 & 7 \end{bmatrix}$$

APPLICATIONS: TOTAL RECYCLE

One of the many tests performed during NASA's Lunar-Mars Life Support Test Project III (LMLSTP III) was the cultivation of wheat to consume CO_2, produce O_2, and contribute to the food requirements of the crew. The gas that flows between the crew's cabin and the plant growth center passes through two concentrators one for oxygen and the other for carbon dioxide:

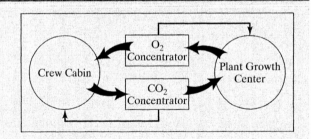

Let us take a look at how Mathcad's matrix-handling features can help us determine the flow rates through the plant growth center and the O_2 concentrator. In the following figure, the unknown streams have been labeled S_1 through S_3.

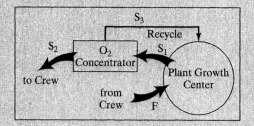

The feed to the plant growth center, F, will be treated as a known commodity. NASA's website[1] states that the concentration of CO_2 in stream F is 85 to 95% pure CO_2. (We'll assume 85% on a molar basis.) The amount of CO_2 entering the plant growth center must be equal to the amount of CO_2 being generated by the crew. The LML-STP III had a crew of four, and a typical rate of CO_2 production for a resting person is 200 ml/min.[2]

With the understanding that the crew is working not resting, we'll assume a CO_2 generation rate of 1000 ml/min. This is the rate at which CO_2 enters the plant growth center, and it is 85% of F. The remainder of F is assumed to contain O_2 and N_2 in the ratio commonly found in air: 21:79 (21 moles of O_2 for every 79 moles of N_2). With these (numerous) assumptions, the contents of stream F can be summarized as follows:

F Contains:	0.0444 moles (per minute)
mole fractions:	0.8500 or 85 mole % CO_2
	0.0313 or 3.13 mole % O_2
	0.1187 or 11.87 mole % N_2

From other statements in the NASA website and a good health assumption about how the O_2 concentrator works, the compositions of the other streams are expected to be something like the following:

Mole Fraction	S_1	S_2	S_3
CO_2	0.00079	0	0.00080
O_2	0.20983	0.85000	0.20665
N_2	0.78938	0.15000	0.79255

---
[1] <http//advlifesupport.jsc.nasa.gov/>
[2] This value is from the "standard man data." in R. C. Seagrave *Biomedical Applications of Heat and Mass Transfer*, Iowa State (Ames, IA, Iowa State University Press, 1971).

With these compositions, we can write material balances to find the molar flow rates in streams S_1 through S_3.

Since CO_2 is being consumed and O_2 is being produced in the growth center, the simplest balance we can write is on N_2. (No reaction term is needed.) The N_2 balance around the plant growth center is

$$N_2 \text{ in } F + N_2 \text{ in } S_3 = N_2 \text{ in } S_1$$

or

$$0.1187 \cdot 0.0444 \text{ mole} + 0.79255 \cdot S_3 = 0.78938 \cdot S_1$$

There is no reaction in the O_2 concentrator, so simple material balances could be written on any component. However, the very small amounts of CO_2 and the very small change in CO_2 composition across the O_2 concentrator make a CO_2 balance a poor choice, because these low-precision values can turn into large errors. So we write O_2 and N_2 balances around the concentrator. The O_2 balance around the O_2 concentrator is

$$O_2 \text{ in } S_1 = O_2 \text{ in } S_3 - O_2 \text{ in } S_3$$

or

$$0.20983 \cdot S_1 = 0.20665 \cdot S_3 - 0.85 \cdot S_2$$

The N_2 balance around the O_2 concentrator is

$$N_2 \text{ in } S_1 = N_2 \text{ in } S_3 - N_2 \text{ in } S_2 2$$

or

$$0.78938 \cdot S_1 = 0.79255 \cdot S_3 - 0.15 \cdot S_2$$

In matrix form, the coefficient matrix and right-hand-side vector for these three equations look like this:

$$F := 0.0444 \cdot \text{mole}$$

$$C := \begin{bmatrix} 0.78938 & -0.15 & -0.79255 \\ -0.78938 & 0 & 0.79255 \\ 0.20983 & -0.85 & -0.20665 \end{bmatrix}$$

$$r := \begin{bmatrix} 0 \\ -0.1187 \cdot F \\ 0 \end{bmatrix} \cdot \text{mol}$$

The coefficient matrix C can then be inverted and multiplied by the right-hand-side vector r to find the flow rates in the three unknown streams:

$$C^{-1} = \begin{bmatrix} -1.414 \cdot 10^3 & -1.349 \cdot 10^3 & 249.591 \\ -6.667 & -6.667 & 0 \\ -1.409 \cdot 10^3 & -1.343 \cdot 10^3 & 248.593 \end{bmatrix}$$

So the flow rates are $S_1 = 7.111$ mole (per minute), $S_2 = 0.035$ mole, and $S_3 = 7.076$ moles.

$$S := C^{-1} \cdot r \quad \begin{bmatrix} 7.111 \\ 0.035 \\ 7.076 \end{bmatrix} \cdot mol$$

SUMMARY

In this chapter, we learned about how to work with matrices in Mathcad. Matrix values can be entered from the keyboard, computed from equations, read from data files, or copied from other programs. Once a matrix exists, there are a variety of ways it can be modified and manipulated to insert or delete rows or columns or to choose a portion of the matrix and assign it to another variable.

The standard array operations were described, including addition, multiplication, transposition, and inversion. The way Mathcad handles these operations is summarized in the following two lists:

MATHCAD SUMMARY

Matrix Fundamentals

ORIGIN	Changes the starting value of the first array element (0 by default).
[Ctrl-M]	Opens the matrix dialog to create a matrix and to insert or delete rows or columns.
Matrix(r,c,f)	Creates a matrix with r rows and c columns, using function f. Function f is user defined and must be a function of two variables.
Identity(c)	Creates an identity matrix with c rows and columns.
[Ctrl-6]	Selects a single column of a matrix.
Submatrix(A, r_{start}, r_{stop}, c_{start}, c_{stop})	Extracts a portion of matrix A.
Augment(A_1, A_2)	Combines arrays A_1 and A_2 side by side.
Stack(A_1, A_2)	Stacks array A_1 on top of array A_2.
READPRN(path)	Reads array values from a text file.
WRITEPRN(path)	Writes array values to a text file.

Matrix Operations

+	Addition of matrices.
[Shift-8]	Matrix multiplication.
→	Vectorize (from the Matrix Toolbar)—used when you want element-by-element multiplication instead of matrix multiplication.
T	Transpose (from the Matrix Toolbar).
[Shift-6]	Invert a matrix.
|M|	Determinant (from the Matrix Toolbar).
sort(v)	Sorts vector v into ascending order.

`reverse(v)`	Reverses the order of vector v—this operation is used after `sort(v)` to get a vector sorted into descending order.
`csort(A,n)`	Sorts array A so that the values in column n are in ascending order.
`rsort(A,n)`	Sorts array A so that the values in row n are in ascending order.

KEY TERMS

array
array origi
augment
column operator
component
determinant
element

element-by-element multiplication
identity matrix
index
input table
invert
matrix

matrix multiplication
range variable
stack
transpose
vector
vectorize

Problems

1. **MATRIX OPERATIONS**

   Given the matrices and vectors

   $$I = \begin{bmatrix} 1 & 0 & 0 & 0 \\ 0 & 1 & 0 & 0 \\ 0 & 0 & 1 & 0 \\ 0 & 0 & 0 & 1 \end{bmatrix}, a = \begin{bmatrix} 1 \\ 2 \\ 3 \\ 4 \end{bmatrix}, b = [2 \ 4 \ 6 \ 8], \text{ and } C = \begin{bmatrix} 2 & 3 & 7 & 11 \\ 1 & 4 & 3 & 9 \\ 0 & 6 & 5 & 1 \\ 1 & 8 & 4 & 2 \end{bmatrix},$$

   which of the following matrix operations are allowed?

   a. I^T — transpose the identity matrix
   b. $|a|$ — find the determinant of vector a
   c. a^{-1} — invert the a vector
   d. $|C|$ — find the determinant of matrix C
   e. C^{-1} — invert the C matrix
   f. $I \cdot a$ — multiply the identity matrix by vector a
   g. $a \cdot b$ — multiply the a vector by vector b
   h. $b \cdot a$ — multiply the b vector by vector a
   i. $C^{-1} \cdot a$ — multiply the inverse of the C matrix by vector a

   For each operation that can be performed, what is the result?

2. **SIMULTANEOUS EQUATIONS**

   The arrays that follow represent coefficient matrices and right-hand-side vectors for sets of simultaneous linear equations written

   $$[C][x] = [r]$$

   in matrix form. Calculate the determinant of these coefficient matrices to see whether each set of simultaneous equations can be solved, and if so, solve for the solution vector [x]:

$$C = \begin{bmatrix} 3 & 1 & 5 \\ 2 & 3 & -1 \\ -1 & 4 & 0 \end{bmatrix}, \quad r = \begin{bmatrix} 20 \\ 5 \\ 7 \end{bmatrix};$$

$$C = \begin{bmatrix} 4 & 2 & 1 \\ 2 & 3 & 0 \\ 0 & 4 & -1 \end{bmatrix}, \quad r = \begin{bmatrix} 18 \\ 6 \\ -2 \end{bmatrix};$$

$$C = \begin{bmatrix} 7 & 3 & 1 \\ 2 & -5 & 6 \\ 1 & 5 & 1 \end{bmatrix}, \quad r = \begin{bmatrix} 108 \\ -62 \\ 56 \end{bmatrix};$$

$$C = \begin{bmatrix} 4 & 2 & 1 & 0 \\ 2 & 3 & 0 & 1 \\ 0 & 4 & -1 & 3 \\ 2 & 1 & 4 & 2 \end{bmatrix}, \quad r = \begin{bmatrix} 13 \\ 8 \\ 4 \\ 19 \end{bmatrix}.$$

3. **SIMULTANEOUS EQUATIONS, II**

   Write the following sets of simultaneous equations in matrix form, and solve (if possible):

   a. $3x_1 + 1x_2 + 5x_3 = 20$
      $2x_1 + 3x_2 - 1x_3 = 5$
      $-1x_1 + 4x_2 = 7$

   b. $6x_1 + 2x_2 + 8x_3 = 14$
      $x_1 + 3x_2 + 4x_3 = 5$
      $5x_1 + 6x_2 + 2x_3 = 7$

   c. $4y_1 + 2y_2 + 1y_3 + 5y_4 = 52.9$
      $3y_1 + y_2 + 4y_3 + 7y_4 = 74.2$
      $2y_1 + 3y_2 + y_3 + 6y_4 = 58.3$
      $3y_1 + y_2 + y_3 + 3y_4 = 34.2$

4. **ELEMENT-BY-ELEMENT MATRIX MATHEMATICS**

   In the following heat exchanger, cold fluid flows through the inside tube and is warmed from T_{C_in} to T_{C_out} by energy from the hot fluid surrounding the inside tube:

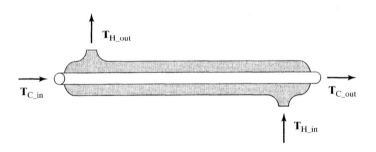

The temperature change of the cold fluid depends on the amount of energy the fluid picks up during the time it is flowing through the exchanger—that is, the rate at which energy is acquired by the cold fluid. The energy and the change in temperature are, respectively,

$$q_{COLD} = \dot{m}\, C_p\, \Delta T_{COLD}$$

and

$$\Delta T_{COLD} = T_{C\_out} - T_{C\_in}.$$

If you performed a series of experiments on this heat exchanger, you might vary the flow rate of the cold fluid to determine the effect on the energy acquired by the cold fluid. Suppose you did this and got the following results:

EXPERIMENT	COLD FLUID RATE kg/minute	HEAT CAPACITY joule/kg K	T_{C_in} °C	T_{C_out} °C	q_{COLD} watts
1	2	4187	6	62	
2	5	4187	6	43	
3	10	4187	6	26	
4	15	4187	6	20	
5	20	4187	6	14	

You could solve for the five q_{COLD} values one at a time, but if you create vectors containing the cold fluid rate values and the ΔT_{COLD} values, Mathcad can solve for all of the q_{COLD} values at once by using element-by-element matrix mathematics.

Create two matrices like the following:

$$\text{flow} := \begin{bmatrix} 2 \\ 5 \\ 10 \\ 15 \\ 20 \end{bmatrix} \cdot \frac{\text{kg}}{\text{min}} \qquad \Delta T_{COLD} := \begin{bmatrix} 62 - 6 \\ 43 - 6 \\ 26 - 6 \\ 20 - 6 \\ 14 - 6 \end{bmatrix} \cdot K$$

Then solve for all five Q_{COLD} values, using element-by-element matrix mathematics. The resulting equation looks like this:

$$q_{COLD} := \overrightarrow{\text{flow} \cdot C_p \cdot \Delta T_{COLD}}$$

The arrow over the right side of the equation is the *vectorize* operator (from the Matrix Toolbar). It is used to tell Mathcad to multiply element by element and is required for this problem.

Note: Degrees Celsius is not a defined unit in Mathcad, but a temperature change of one degree Celsius is equal to a temperature change of one degree on the kelvin scale—so kelvins can be specified for the ΔT_{COLD} matrix.

5. ## ELEMENT-BY-ELEMENT MATRIX MATHEMATICS

The preceding problem considered only the energy acquired by the cold fluid as it passed through the heat exchanger. This problem considers the energy transferred from the hot fluid to the cold fluid (the energy passing across the tube of the heat exchanger.) The diagram is the same as before:

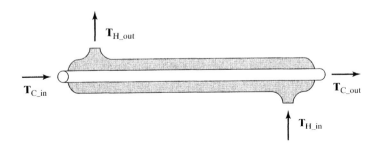

The rate at which energy crosses a heat exchanger tube is predicted with the use of a heat transfer coefficient h. The equation is

$$q_{HX} = hA\Delta T_{LM}$$

where A is the area through which the energy passes (the surface area of the tube) and ΔT_{LM} is the *log mean temperature difference*, defined, for the heat exchanger shown, as

$$\Delta T_{LM} = \frac{(T_{H\_out} - T_{C\_in}) - (T_{H\_in} - T_{C\_out})}{\ln\left[\frac{(T_{H\_out} - T_{C\_in})}{(T_{H\_in} - T_{C\_out})}\right]} = \frac{\Delta T_{left} - \Delta T_{right}}{\ln\left[\frac{\Delta T_{left}}{\Delta T_{right}}\right]}$$

Given the following data, use element-by-element matrix mathematics to determine all of the q_{HX} values with one calculation:

EXPERIMENT	H	A	T_{C_IN}	T_{C_OUT}	T_{H_IN}	T_{H_OUT}	q_{HX}
	W/m² K	m²	°C	°C	°C	°C	watts
1	300	0.376	6	36	75	54	
2	450	0.376	6	48	73	51	
3	600	0.376	6	52	71	47	
4	730	0.376	6	57	77	40	
5	1200	0.376	6	60	74	35	

Note: The natural logarithm is not defined for matrices, so element-by-element mathematics is required for this problem. Use the *vectorize* operator (from the Matrix Toolbar) over the entire right side of defining equations to tell Mathcad to use element-by-element mathematics.

6. MATERIAL BALANCES ON A GAS ABSORBER

The equations that follow are material balances for CO_2, SO_2, and N_2 around the gas absorber shown in the accompanying figure. Stream S_1 is known to contain 99 mole % MEA and 1 mole % CO_2. The flow rate in S_1 is 100 moles per minute.

The compositions used in the material balances are tabulated as follows:

COMPONENT	S_1	S_2	S_3	S_4
CO_2	0.01000	0.07522	0.08000	0.00880
SO_2	0	0.01651	0.02000	0.00220
N_2	0	0	0.90000	0.98900
MEA	0.99000	0.90800	0	0

CO_2 Balance: CO_2 in S_1 + CO_2 in S_3 = CO_2 in S_2 + CO_2 in S_4
1 mole + $0.08000 \cdot S_3 = 0.07522 \cdot S_2 + 0.00880 \cdot S_4$

SO_2 Balance: SO_2 in S_1 + SO_2 in S_3 = SO_2 in S_2 + SO_2 in S_4
$0 + 0.02000 \cdot S_3 = 0.01651 \cdot S_2 + 0.00220 \cdot S_4$

N_2 Balance: N_2 in S_1 + N_2 in S_3 = N_2 in S_2 + N_2 in S_4
$0 + 0.90000 \cdot S_3 = 0 + 0.98900 \cdot S_4$

Solve the material balances for the unknown flow rates S_2 through S_4.

7. MATERIAL BALANCES ON AN EXTRACTOR

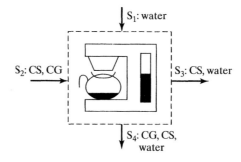

This problem focuses on a low-cost, high-performance chemical extraction unit: a drip coffeemaker. The ingredients are water, coffee solubles (CS), and coffee grounds (CG). Stream S_1 is water only, and the coffeemaker is designed to hold 1 liter. Stream S_2 is the dry coffee placed in the filter and contains 99% grounds and 1% soluble ingredients. The product coffee contains 0.4% CS and 99.6% water. Finally, the waste product (S_3) contains 80% CG, 19.6% water, and 0.4% CS. (All percentages are on a volume basis.)

Write material balances on water, CS, and CG, and then solve the material balances for the volumes S_2 through S_4.

8. FLASH DISTILLATION

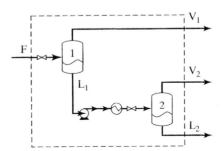

When a hot, pressurized liquid is pumped into a tank (flash unit) at a lower pressure, the liquid boils rapidly. This rapid boiling is called a *flash*. If the liquid contains a mixture of chemicals, the vapor and liquid leaving the flash unit will have different compositions, and the flash unit can be used as a separator. The physical principle involved is vapor–liquid equilibrium: The vapor and the liquid leaving the flash unit are in equilibrium. This allows the composition of the outlet streams to be determined from the operating temperature and pressure of the flash unit. Multiple flash units can be used together to separate multi-component mixtures.

A mixture of methanol, butanol, and ethylene glycol is fed to a flash unit operating at 165°C and 7 atm. The liquid from the first flash unit is recompressed, reheated, and sent to a second flash unit operating at 105°C and 1 atm. The compositions of the feed stream F and the three product streams are listed in the following table:

	MASS FRACTION IN STREAM			
COMPONENT	FEED	V_1	V_2	L_2
Methanol	0.300	0.716	0.533	0.086
Butanol	0.400	0.268	0.443	0.388
Ethylene Glycol	0.300	0.016	0.024	0.526

The mixture is fed to the process at a rate of 10,000 kg/h. Write material balances for each chemical, and then solve for the mass flow rate of each product stream (V_1, V_2, L_2). A material balance is simply a mathematical statement that all of the methanol (for example) going into the process has to come out again (assuming a steady state and no chemical reactions.) The methanol balance is as follows:

methanol in the feed = methanol in V_1 + methanol in V_2 + methanol in L_2;

$0.300 \cdot (10{,}000 \text{ kg/h}) = 0.716 \cdot V_1 + 0.533 \cdot V_2 + 0.086 \cdot L_2$.

GRADES
87
85
43
62
97
88
87
58
67
79
91
82
80
73
58

9. **FINDING THE MEDIAN GRADE**

It is common for instructors to report the average grade on an examination. An alternative way to let students know how their scores compare with the rest of the class is to find the *median* score. If the scores are sorted, the median score is the value in the middle of the data set.

a. Use the sort( ) function to sort the 15 scores (elements 0 to 14) listed at the left, and then display the value of element 7 in the sorted array. Element 7 will hold the median score for the sorted grades.

b. Check your result by using Mathcad's median( ) function on the original vector of grades.

10. **CURRENTS IN MULTILOOP CIRCUITS**

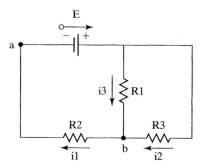

Multiloop circuits are analyzed using Kirchhoff's voltage law, described in Chapter 2, and Kirchhoff's current law, namely,

At any junction in a circuit, the input current(s) must equal the output current(s).

The latter is simply a statement of conservation of current: All of the electrons entering a junction have to leave again, at least at steady state.

Applying the current law at point *b* in the accompanying diagram yields

$$i_3 + i_2 = i_1.$$

Applying the voltage law to the left loop and the overall loop provides two more equations:

$$E - i_3 R_1 - i_1 R_2 = 0;$$
$$E - i_2 R_3 - i_1 R_2 = 0.$$

We thus have three equations in the unknowns i_1, i_2, i_3.

Use the following resistance values, and solve the system of simultaneous equations to determine the current in each portion of the circuit:

$$E = 9 \text{ volts};$$
$$R_1 = 20 \text{ ohms};$$
$$R_2 = 30 \text{ ohms};$$
$$R_3 = 40 \text{ ohms}.$$

11. **THE WHEATSTONE BRIDGE**

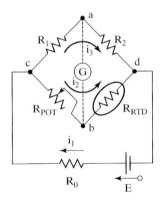

Resistance temperature devices (RTDs) are often used as temperature sensors. As the temperature of the device changes, its resistance changes. If you can determine an RTD's resistance, you can look on a table to find the temperature. A circuit known as a *Wheatstone bridge* can be used to measure unknown resistances. In the preceding figure, an RTD has been built into the bridge as the unknown resistance.

To use the Wheatstone bridge, the adjustable resistor R_{POT} (called a potentiometer) is adjusted until the galvanometer G shows that points a and b are at the same potential. Once the bridge has been adjusted so that no current is flowing through the galvanometer (because there is no difference in potential between a and b), the reading on the potentiometer and the known resistances R_1 and R_2 can be used to compute the resistance of the RTD.

How Does It Work?

Because resistors R_1 and R_{POT} come together at point c, and we know that the voltages at a and b are the same, there must be the same voltage drop across R_1 and R_{POT} (not the same current). Thus, we have

$$i_3 \cdot R_1 = i_2 \cdot R_{POT}.$$

Similarly, because R_2 and R_{RTD} are connected at point d, we can say that

$$i_3 \cdot R_2 = i_2 \cdot R_{RDT}.$$

If you solve for i_3 in one equation and substitute into the other, you get

$$R_{RDT} = R_{POT} \frac{R_2}{R_1}.$$

In this way, if you know R_1 and R_2 and have a reading on the potentiometer, you can calculate R_{RDT}.

a. Given the following resistances, what is R_{RDT}? Assume that a 9-volt battery is used for E.

$$R_0 = 20 \text{ ohms};$$
$$R_1 = 10 \text{ ohms};$$
$$R_2 = 5 \text{ ohms};$$
$$R_{POT} \text{ adjusted to } 12.3 \text{ ohms}.$$

b. Use Kirchhoff's current law at either point c or d, and Kirchhoff's voltage law to determine the values of i_1, i_2, and i_3.

12. STEADY-STATE CONDUCTION, I

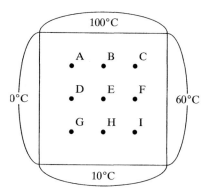

Steady-state conduction in two dimensions is described by Laplace's equation:

$$\frac{\partial^2 T}{\partial x^2} + \frac{\partial^2 T}{\partial y^2} = 0.$$

The partial derivatives in Laplace's equation can be replaced by finite differences (see Chapter 7) to obtain approximate solutions giving the expected temperature at various points in the two-dimensional region. The difference equation is

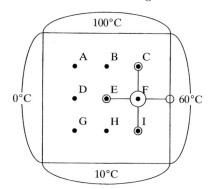

$$\frac{T_{i-1,j} - 2T_{i,j} + T_{i+1,j}}{(\Delta x)^2} + \frac{T_{i,j-1} - 2T_{i,j} + T_{i,j+1}}{(\Delta y)^2} = 0,$$

where $T_{i,j}$ is the temperature at a point (i, j) in the conducting region, $T_{i-1,j}$ is the temperature at the point to the left of point i, j, etc. If Δx and Δy are equal, these two terms can be combined and the subscripts replaced by more descriptive terms, yielding

$$T_{\text{left}} + T_{\text{above}} + T_{\text{right}} + T_{\text{below}} - 4T_{\text{center}} = 0$$

Using this equation at each of the nine points in the region shown in the accompanying diagram generates nine equations in nine unknowns that can be solved simultaneously for the temperature at each point. For example, applying the equation at point F gives the equation

$$T_E + T_C + 60° + T_I - 4T_F = 0.$$

Develop the system of nine equations, write them in matrix form, and determine the temperature at points A through I.

13. STEADY-STATE CONDUCTION, II

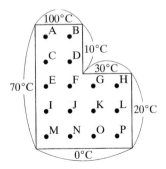

In the previous problem, Laplace's equation was applied to steady-state conduction in a square region. Actually, the equation can be applied to any region composed of rectangles or any shape that can be approximated with rectangles. Again, the final version of the equation presented in the last problem does require that the distance between the points be the same in the x and y directions.

Apply Laplace's equation to the foregoing L-shaped region, and determine the temperatures at points A through P.

Note: Solving for all 16 temperatures simultaneously requires a 16×16 matrix (256 elements). Using [Ctrl-m] to create a matrix will restrict you to a maximum of 100 elements. There are several ways to work around this limitation, such as using `stack()` or `augment()` functions, or copying and pasting the array from a spreadsheet.

14. TRAFFIC FLOW, part I

A traffic count has been performed on a series of one-way streets. After one hour of counting, the counts shown on the following figure were recorded:

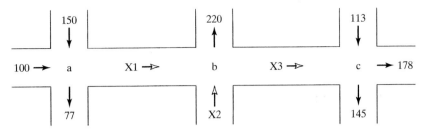

Since all of the vehicles that enter an intersection must leave again, we can write total vehicle balances on each intersection:

Intersection a:	$100 + 150 = X1 + 77$	or	$X1 = 173;$
Intersection b:	$X1 + X2 = 220 + X3$	or	$X1 + X2 - X3 = 220;$
Intersection c:	$X3 + 113 = 178 + 145$	or	$X3 = 210.$

These balances at the intersections represent three equations in three unknowns and can be solved using matrix methods. (Admittedly, this simple problem hardly needs to be solved using such methods.)

a. What would the coefficient matrix and right-hand-side vector look like for the preceding system of three equations in three unknowns?

b. Solve the three equations simultaneously using matrix methods.

15. TRAFFIC FLOW, part II

    Since the number of vehicles entering and leaving an intersection must be equal, the four intersections in the following road pattern allow four vehicle balances to be written that can be solved for four unknown vehicle counts.

    a. Write vehicle balances for each of the four intersections.
    b. Solve for the four unknown traffic counts using matrix methods.
    c. Is there any evidence that people are making U-turns in the intersections?

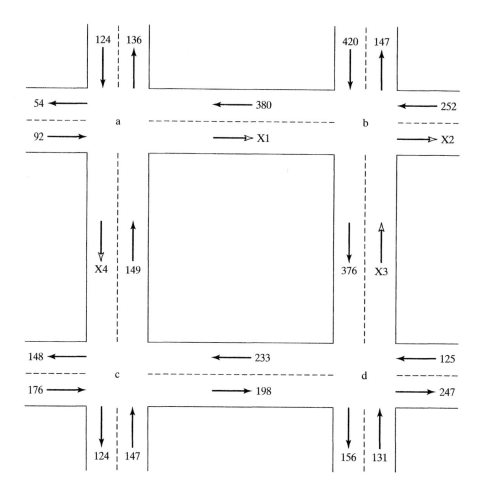

5
Data Analysis Functions

RISK ANALYSIS

Each year, we read in the newspapers of earthquakes with terrible consequences in terms of lost lives and damaged property. There have also been serious earthquakes in highly populated areas in California, with, relatively speaking, considerably less damage. The difference is due to *risk analysis* and *risk management*. Because of the serious threat earthquakes pose in many parts of California, the state has spent a great deal of time and effort (as well as money) trying to learn what might happen in an earthquake (risk analysis) and has invested heavily in trying to minimize the consequences of a major earthquake (risk management).

Because engineers work on projects that can affect large numbers of people, safety and risk management are a standard practice. What happens if a device fails? If a power line fails, can we get power to the hospital from another direction? Can we use an inflatable bag to improve passenger safety in an automobile accident? But what about the safety of small children in automobiles with air bags? These questions are all related to risk analysis and management.

Understanding the potential risks associated with a process or product requires a thorough knowledge of the process or product, as well as the ability and willingness to consider what might go wrong.

SECTIONS

5.1 Graphing with Mathcad
5.2 Statistical Functions
5.3 Interpolation
5.4 Curve Fitting

OBJECTIVES

After reading this chapter, you will
- be able to create graphs in Mathcad
- know how to use a QuickPlot to see what a function looks like when plotted
- know that Mathcad provides built-in functions for basic statistical calculations
- be able to use Mathcad's linear and spline interpolation functions on data sets
- know two ways to perform linear regression on data sets

It takes a lot of work to anticipate how a process or product might fail and to quantify the outcome of a potential failure. There are models to help quantify risk, but they require good data on the products and processes involved. Getting the data into a form that the models can use requires a lot of analysis. That's why risk analysis is a good lead-in to a chapter on Mathcad's data analysis functions.

5.1 GRAPHING WITH MATHCAD

Engineers spend a lot of time analyzing data. By looking at data from a current design, you can learn what can be changed to make the device, system, or process work better. Mathcad's data analysis functions can provide a lot of help in understanding and evaluating data.

One of the most fundamental steps in analyzing data is visualizing the data, usually in the form of a graph. Mathcad provides a variety of graphing options.

Plotting Vector against Vector

The easiest way to create a graph in Mathcad is simply to plot one vector against another. For example, a vector of temperature values can be plotted against a vector of time values. This method works well for many data analysis situations, because the data are often already available in vector form, as in the following example:

Time =

	0
0	0
1	1
2	2
3	3
4	4
5	5
6	6
7	7
8	8
9	9

min   Temp =

	0
0	298
1	299
2	301
3	304
4	306
5	309
6	312
7	316
8	319
9	322

K

When you plot one vector against another, the vectors must each have the same number of elements. The preceding vectors each have 10 elements, so they can be plotted against each other. To create an x–y graph, use the Graph Toolbox and click on the X–Y Plot button, or use the keyboard shortcut [Shift-2]. The result is as follows:

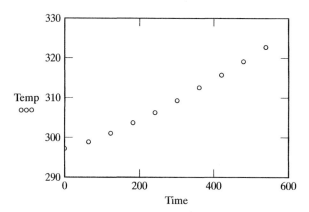

Note that the temperature values look fine, but the time values are not the same as those shown in the Time vector. Mathcad always plots values in its default (base) units. The base unit for temperature (in the default SI unit system) is kelvins, so the temperature values were unaffected. The time values were plotted as Mathcad saved them, in seconds. There is a trick that is sometimes used to get the right values to display on the graph: You can get the values in minutes by dividing the Time on the x–axis by minutes, as shown in the following figure:

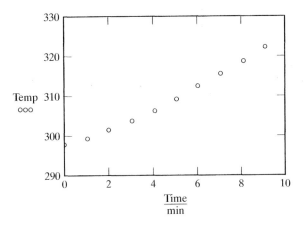

Now the right values are displayed, but the axis labels are hard to interpret. An alternative is to remove the units from the Time and Temp vectors and then plot the values. Then you can add your own axis labels to indicate the units. (Double-click on the graph to add axis labels.) Note that removing the units from the Temp values is not really necessary in this example, but it was done to demonstrate a general approach to graphing values with units. Using this approach, we obtain

$$\text{Time}_{plot} := \frac{\text{Time}}{\text{min}}$$

$$\text{Temp}_{plot} := \frac{\text{Temp}}{\text{K}}$$

The plotted result is

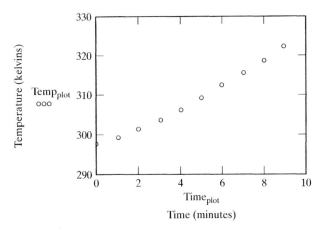

The process of removing the units and plotting only values is exactly what is being accomplished with the "trick" used to plot time on the previous graph. Hopefully, making the process more explicit and including units with the axis labels makes the graph easier to read.

Plotting Multiple Curves

Mathcad allows up to 16 curves on a single graph. If you have two vectors of *y* values corresponding to the same set of *x* values, you plot the second curve to the graph as follows:

$$
\text{Time} = \begin{array}{|c|c|} \hline & 0 \\ \hline 0 & 0 \\ 1 & 1 \\ 2 & 2 \\ 3 & 3 \\ 4 & 4 \\ 5 & 5 \\ 6 & 6 \\ 7 & 7 \\ 8 & 8 \\ 9 & 9 \\ \hline \end{array} \text{min} \quad \text{Temp} = \begin{array}{|c|c|} \hline & 0 \\ \hline 0 & 298 \\ 1 & 299 \\ 2 & 301 \\ 3 & 304 \\ 4 & 306 \\ 5 & 309 \\ 6 & 312 \\ 7 & 316 \\ 8 & 319 \\ 9 & 322 \\ \hline \end{array} \text{K} \quad \text{Temp2} = \begin{array}{|c|c|} \hline & 0 \\ \hline 0 & 305 \\ 1 & 307 \\ 2 & 311 \\ 3 & 316 \\ 4 & 322 \\ 5 & 330 \\ 6 & 341 \\ 7 & 354 \\ 8 & 367 \\ 9 & 382 \\ \hline \end{array} \text{K}
$$

1. Click on the placeholder on the *y*-axis.
2. Press [Comma]. This causes Mathcad to create a second placeholder on the *y*-axis:

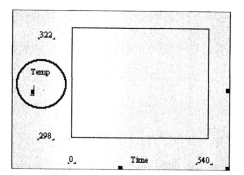

3. Enter the names of the vectors of *y* values in the placeholders on the *y*-axis:

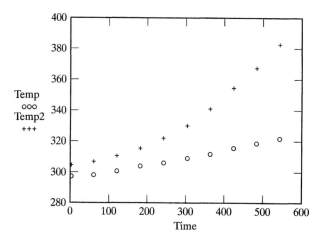

If the vectors you want to plot do not share the same x values, you need to have two vectors indicated on each axis. For example, the following x vectors contain different values, but span nearly the same range, and each x vector has a corresponding y vector (x1 and y1 are related, as are x2 and y2):

$$x1 := \begin{pmatrix} 1 \\ 3 \\ 5 \\ 7 \\ 9 \end{pmatrix} \quad y1 := \begin{pmatrix} 2 \\ 5 \\ 9 \\ 16 \\ 28 \end{pmatrix} \quad x2 := \begin{pmatrix} 1 \\ 2 \\ 3 \\ 4 \\ 5 \\ 6 \\ 7 \end{pmatrix} \quad y2 := \begin{pmatrix} 3 \\ 6 \\ 8 \\ 13 \\ 20 \\ 27 \\ 35 \end{pmatrix}$$

These vectors can be plotted on the same graph by putting two placeholders on each axis. The extra placeholders are created by clicking on the existing placeholders and then pressing [Comma]. (This must be done on both the x-axis and the y-axis.) The result is

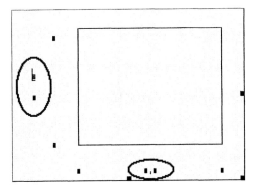

Then fill in the placeholders with the vector names to create the graph:

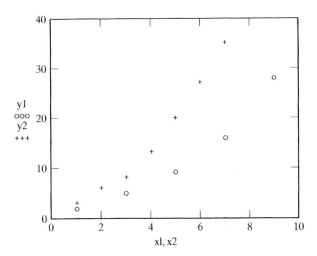

Element-by-Element Plotting

In older versions of Mathcad, the standard method for graphing was to plot one element of one vector against one element of another vector. This method is still available and can sometimes be convenient. To plot the temperature and time vectors element by element, we must define a *range variable* containing as many elements as each of the vectors. This is easily done using the `last(v)` function as follows:

$$i := 0 \; .. \; \text{last(Time)}$$

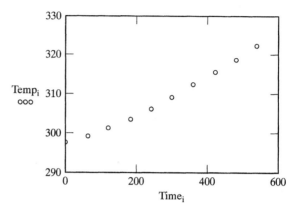

The index of the last element of the `Time` vector was 9, so the range variable, `i`, was defined over the range 0 to 9. The range indicator ("..", or ellipsis) is entered by using the [;] (semicolon) key. The subscripts on the `Temp` and `Time` vectors are array index subscripts, entered by using the [[ ] (left square bracket) key. While this plot looks the same as that created by plotting the vectors directly, the use of the range variable gives you a lot of control over how the elements are plotted. For example, you could change the order of the plotted elements by changing the subscript on the `Temp` vector. The graph will then look like this:

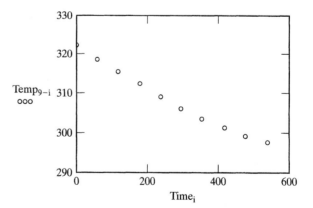

QuickPlots

The final method for creating a graph using Mathcad is a QuickPlot, which is used when you want to see what a function looks like. You start by creating an *x-y* graph and then entering the function on the *y*-axis. Functions always have one or more parameters, such as the `x` in the function `sin(x)`. Next, enter the parameter on the *x*-axis. By default, Mathcad will automatically evaluate the function for a range of parameter values, from −10 to +10. For `sin(x)`, the results will look like this:

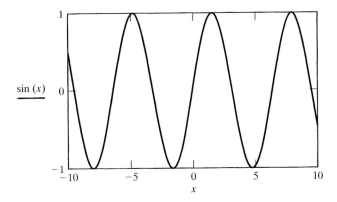

To plot a different range, simply click on the x-axis of the graph and change the displayed axis limits:

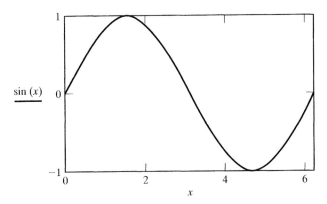

Here, the x–axis limits were changed to 0 and 2π to plot one complete sine wave.

PRACTICE!

a. Try QuickPlots of other common functions, like ln(x) and exp(x).
b. See what Quickplots of the following functions look like (the Φ function is available on the Greek Toolbar, the absolute value operator on the Calculator Toolbar):
   - if (sin(x) > 0, −sin(x), sin(x))
   - $\Phi(|\text{mod}(x, 2)| - 1)$—this uses the Heaviside step function Φ to generate a square wave

Modifying Graphical Display Attributes

You can change the appearance of a graph by double-clicking anywhere on the graph. This will bring up the graph-formatting dialog box. From this dialog box, you can change the

- axis characteristics (linear or logarithmic plots, for example, and whether grid lines are displayed),
- trace (curve) characteristics, such as the symbol and line style, and
- text and position of labels.

For example, we can use the formatting dialog box to show a curve by using a dashed line and adding grid lines and labels to the sine graph. The necessary changes to the dialog boxes, are circled in the figures that follow.

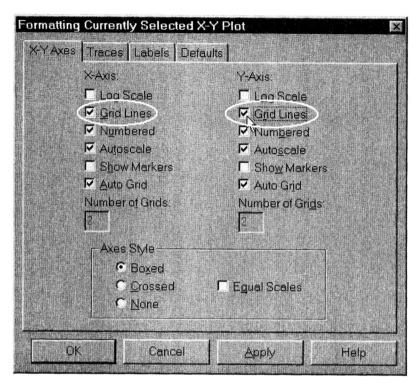

To change a graph's format, double-click in the middle of the graph to open the formatting dialog box. Then turn on the grid lines by using the check boxes on the X–Y Axes panel. Next, switch to the Traces panel and choose a dashed line for trace 1 (there is only one trace, or curve, displayed on this graph):

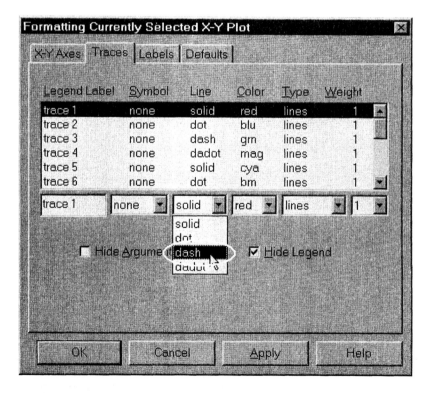

Finally, add a title and axis labels to the graph by using the Labels panel. The screen should look like this:

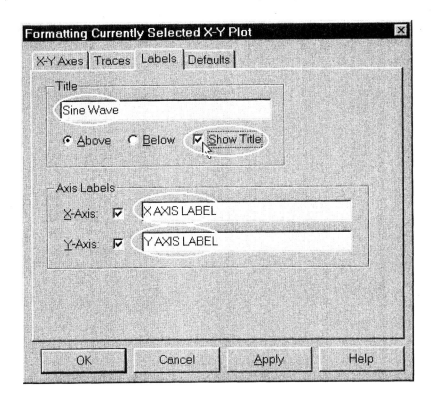

Then, click on the dialog's OK button to make the changes to the graph. The graph will look like this:

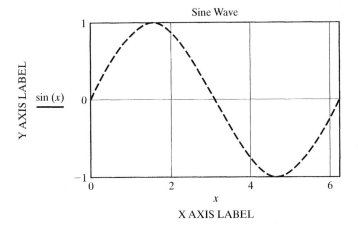

5.2 STATISTICAL FUNCTIONS

Mathcad provides several functions for commonly used statistical calculations, including mean(A), stdev(A), and var(A) to calculate, respectively, the mean, standard deviation, and variance of a population. As the following examples illustrate, these functions work with vectors or matrices, and they can handle units:

APPLICATION: SAMPLES AND POPULATIONS

- The transparency of each windshield on an assembly line might be measured before the windshield is installed. Since every windshield is included in the data set, the data set is called a *population*.
- One out of every thousand windshields might be tested for impact strength (i.e., it might be broken). The results from the broken windshields are used to represent the impact strength of all the windshields. Since the data set does not include all of the windshields, it is called a *sample*. Any time you use a sample, you need to be careful to try to get a *representative sample* (usually a random sample).

$$C = \begin{bmatrix} 1 & 1 \\ 2 & 8 \\ 3 & 27 \\ 4 & 64 \\ 5 & 128 \end{bmatrix} \qquad a = \begin{bmatrix} 3 \\ 2 \\ 7 \\ 4 \end{bmatrix}$$

$C_{avg} := \text{mean}(C)$ $\qquad C_{avg} = 24$
$a_{avg} := \text{mean}(a)$ $\qquad a_{avg} = 4$
$\text{Temp}_{avg} := \text{mean}(\text{Temp})$ $\qquad \text{Temp}_{avg} = 308.733 \cdot K$
$\text{Temp}_{std} := \text{stdev}(\text{Temp})$ $\qquad \text{Temp}_{std} = 7.961 \cdot K$
$\text{Temp}_{var} := \text{var}(\text{Temp})$ $\qquad \text{Temp}_{var} = 63.377 \cdot K^2$

The functions for the standard deviation and variance of a sample are `Stdev(A)` and `Var(A)`, respectively. The uppercase `S` and `V` are used to differentiate these functions for samples from those for populations.

PRACTICE!

The standard deviation lets you know how far a typical value is from the mean value of the data set. The data sets shown here have approximately the same mean value:

$$\text{Set}_1 = \begin{bmatrix} 2.5 \\ 2.5 \\ 2.5 \\ 2.5 \\ 2.5 \end{bmatrix} \qquad \text{Set}_2 = \begin{bmatrix} 2.2 \\ 2.6 \\ 2.6 \\ 2.5 \\ 2.4 \end{bmatrix} \qquad \text{Set}_3 = \begin{bmatrix} 1.6 \\ 1.6 \\ 2.8 \\ 2.1 \\ 3.8 \end{bmatrix}$$

Which has the greatest standard deviation? Use Mathcad's `stdev( )` function to check your result.

5.3 INTERPOLATION

If you have a data set—for example, a set of temperature values at various times—and you want to predict a temperature at a new time, you could fit a function to the data and calculate the predicted temperature at the new time. Or you can interpolate between data set values. Mathcad provides a number of functions for linear interpolation and cubic spline interpolation. The linear interpolation function, `linterp(vx, vy, x_new)`, predicts a new y value at the new x value (specified in the function's argument list) by using a linear interpolation and by using the x values in the data set nearest to the new x value.

As an example, consider the temperature and time data presented at the beginning of this chapter:

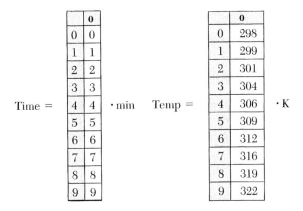

The temperature vector includes temperatures at 2 and 3 minutes, but not at 2.3 minutes. We can use the `linterp( )` function to interpolate between the temperature values at 2 and 3 minutes to predict the temperature at 2.3 minutes:

$$\text{Temp}_{\text{interp}} := \text{linterp (Time, Temp, 2.3} \cdot \text{min)}$$
$$\text{Temp}_{\text{interp}} = 302.165 \cdot \text{K}$$

You can also use `linterp( )` to extrapolate beyond the limits of a data set, although extrapolation is always risky. The previous temperature data were collected from time 0 to 9 minutes. We can extrapolate to a time of 20 minutes by using `linterp( )`:

$$\text{Temp}_{\text{interp}} := \text{linterp (Time, Temp, 20} \cdot \text{min)}$$
$$\text{Temp}_{\text{interp}} = 360.398 \cdot \text{K}$$

Extrapolation will give you a result, but there is no guarantee that it is correct. Since the result is outside the range of the data set, the result is uncertain. For this data set, the heater might have been turned off after 9 minutes, and the temperatures might have started to decrease with time. Because the data set includes only temperatures between 0 and 9 minutes, we cannot know what happened at later times.

An alternative to linear interpolation is *cubic spline interpolation*. This technique puts a curve (a cubic polynomial) through the data points, which results in a curve that passes through each data point with continuous first and second derivatives. Interpolation for values between points can be done using the cubic polynomial. To use cubic spline interpolation in Mathcad, you first need to fit the cubic polynomial to the data by using the `cspline(vx, vy)` function. This function returns a vector of second-derivative values, vs, that is then used in the interpolation, which is performed with the `interp(vs, vx, vy, x_new)` function:

$$\text{vs} := \text{cspline (Time, Temp)}$$
$$\text{Temp}_{\text{interp}} := \text{interp (vs, Time, Temp, 2.3} \cdot \text{min)}$$
$$\text{Temp}_{\text{interp}} = 302.136 \cdot \text{K}$$

You can also extrapolate with the cubic spline, but doing so is risky. Because the cubic spline technique fits three adjacent points with a cubic polynomial, it runs into trouble at each end of the data set, since the first and last points do not have another point on each side. Hence, the cubic spline method must do something special for these endpoints. Mathcad provides three methods for handling the endpoints:

- `cspline( )`—creates a spline curve that is cubic at the endpoints.
- `pspline( )`—creates a spline curve that is parabolic at the endpoints.
- `lspline( )`—creates a spline curve that is linear at the endpoints.

These spline functions will all yield the same interpolated results for interior points, but can give widely varying results if you extrapolate your data set:

```
vs := cspline (Time, Temp)
     interp (vs, Time, Temp, 20·min) = 950 K
vs := pspline (Time, Temp)
     interp (vs, Time, Temp, 20·min) = 384 K
vs := lspline (Time, Temp)
     interp (vs, Time, Temp, 20·min) = 232 K
```

CAUTIONS

1. Extrapolation (predicting values outside the range of the data set) should be avoided whenever possible. The temperatures used in the foregoing example were recorded for times ranging from 0 to 9 minutes. Predicting a temperature at 9.2 minutes is safer than predicting one at 90 or 900 minutes, but you are still making an assumption that nothing changed between 9.0 and 9.2 minutes. (The researcher might have turned off the heater, and the temperatures might have started falling.)

2. Cubic spline fits to noisy data can produce some amazing results. In the accompanying graph, an outlier at $x = 5$ was included in the data set to demonstrate how this single bad point affects the spline curve for adjacent points as well. Interpolation using these curves is a very bad idea. Regression to find the "best fit" curve and then using the regression result to predict new values is a better idea for noisy data.

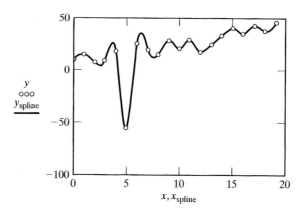

Using a QuickPlot to Plot the Spline Curve

If you want to see what the spline curve looks like, Mathcad's QuickPlot feature will allow you to see the curve with little effort. Since a QuickPlot will evaluate a function multiple times over a range of values, simply let Mathcad evaluate the `interp( )` function many times and display the result. For the temperature–time data presented earlier, this means calculating the second-derivative values by using `cspline( )` (or `pspline( )` or `lspline( )`) and then creating the QuickPlot.

In the figure that follows, the t in the `interp( )` function and on the x-axis is a dummy variable. Mathcad evaluates the `interp( )` function for values of t between −10 and 10 (by default) and displays the result. We need to change the limits on the x-axis to coincide with the time values in the data set, 0 to 9 minutes. To change the axis limits, click on the x-axis and then edit the limit values:

```
vs := cspline (Time, Temp)
```

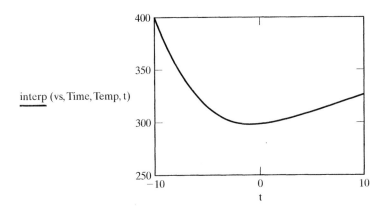

Note: The graph also demonstrates why it is not a good idea to extrapolate by using the `interp( )` function. The data set gives no evidence that the temperatures were very high before the experiment started (negative times), but that is what the `cspline( )` fit predicts. The results obtained by using `lspline( )` would be very different. (Check them out!)

The spline fit of the actual data set (not extrapolated) with the x-axis limits changed is shown in the following graph:

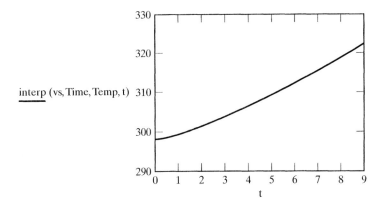

The curve looks well behaved in this range of times, typical of a spline fit to clean data.

PRACTICE!

The following "noisy" data were created by adding random numbers to the "clean" data set

$$x = \begin{bmatrix} 0.0 \\ 0.5 \\ 1.0 \\ 1.5 \\ 2.0 \\ 2.5 \\ 3.0 \end{bmatrix} \quad y_{clean} = \begin{bmatrix} 0.00 \\ 0.48 \\ 0.84 \\ 1.00 \\ 0.91 \\ 0.60 \\ 0.14 \end{bmatrix} \quad y_{noisy} = \begin{bmatrix} 0.24 \\ 0.60 \\ 0.69 \\ 1.17 \\ 0.91 \\ 0.36 \\ 0.18 \end{bmatrix}$$

a. Compare the cubic spline fit of each data set by plotting the cubic spline interpolation for each.
b. Use the `interp( )` function to predict the y-value corresponding to $x = 1.7$ for each data set.

5.4 CURVE FITTING

Simple Linear Regression

Mathcad provides a number of functions to fit curves to data and to use fitted curves to predict new values. One of the most common curve-fitting applications is *linear regression*. Simple linear regression is carried out in Mathcad by using two or three of the following functions: `slope (vx, vy)`, `intercept (vx, vy)`, and `corr (vx, vy)`. These functions respectively return the *slope*, *intercept*, and *correlation coefficient* of the best fit straight line through the data represented by the x and y vectors. For example, for the time–temperature data of the preceding section,

```
b:= intercept (Time, Temp)     b = 296.369·K

m:= slope (Time, Temp)         m = 2.756· K/min
```

The temperature values predicted by the model equation can be calculated from the slope and intercept values by using a range variable to keep track of the times. Then the quality of the fit can be determined by using the `corr( )` function:

```
i := 0 .. last(Time)

Temp_pred_i := b + m·Time_i

R2 := corr(Temp, Temp_pred)²    R2 = 0.98904
```

Here, the correlation coefficient was squared to calculate the *coefficient of determination* (usually called the R^2 value) for the regression. An R^2 value of 1 implies that the regression line is a perfect fit to the data. The value 0.98904 suggests that the regression line is a pretty good fit to the data, but it is always a good idea to plot the original data together with the regression line. The result looks like this:

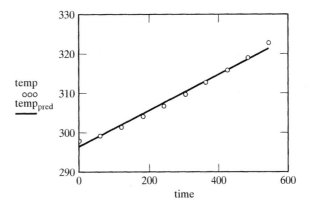

With this plot, we can clearly see that the actual curve of temperature versus time is not linear. This becomes quite apparent if you plot the *residuals*:

```
Residual_i:=Temp_i - Temp_pred_i
```

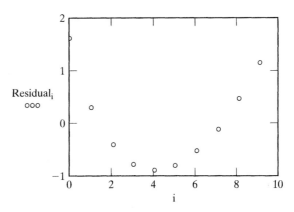

A residual plot should show randomly scattered points. Obvious trends in such a plot, such as the "U" shape in the preceding one, suggest that the equation used to fit the data was a poor choice. Here, it is saying that the linear equation is not a good choice for fitting these nonlinear data.

The foregoing example was included as a reminder that, while you can perform a simple (straight-line) linear regression on any data set, it is not always a good idea to do so. Nonlinear data require a more sophisticated curve-fitting approach.

PRACTICE!

Two sets of y values are shown. The noisy data were calculated from the clean data by adding random values. Try linear regression on these data sets. What is the impact of noisy data on the calculated slope, intercept, and R^2 value?

$$x = \begin{bmatrix} 0 \\ 1 \\ 2 \\ 3 \\ 4 \\ 5 \end{bmatrix} \quad y_{clean} = \begin{bmatrix} 1 \\ 2 \\ 3 \\ 4 \\ 5 \\ 6 \end{bmatrix} \quad y_{noisy} = \begin{bmatrix} 0.85 \\ 1.91 \\ 3.03 \\ 3.96 \\ 5.10 \\ 5.90 \end{bmatrix}$$

124 Chapter 5 Data Analysis Functions

PROFESSIONAL SUCCESS

Visualize your data and results.
Numbers, such as those in the following data set, are a handy way to store information:

X	Y	X	Y
−1.0000	−0.5478	0.1367	0.0004
−0.9900	−0.4252	0.1543	0.3042
−0.9626	−0.3277	0.2837	0.5407
−0.9577	−0.7865	0.4081	0.0122
−0.9111	−0.2327	0.4242	0.7593
−0.8391	−0.9654	0.5403	0.0304
−0.7597	−0.1030	0.6469	0.0565
−0.7481	−0.9997	0.6603	0.9821
−0.6536	−0.9791	0.7539	0.0999
−0.5477	−0.0319	0.8439	0.9617
−0.5328	−0.8888	0.9074	0.2277
−0.4161	−0.0130	0.9147	0.8781
−0.2921	−0.0043	0.9602	0.7795
−0.2752	−0.5260	0.9887	0.4184
−0.1455	−0.0005	0.9912	0.6536
0.0044	0.0088	1.0000	0.5403

For most of us, however, a data set is hard to visualize. A quick glance at the data at the left suggests that the y values get bigger as the x values get bigger. To try to describe these data, linear regression could be performed to calculate a slope and an intercept:

m := slope(x,y)   m = 0.653
b := intercept(x,y)   b = 1.211•10⁻³
r2 := corr(x, y)²   r2 = 0.667

But by graphing the data, it becomes clear that simple linear regression is not appropriate for this data set:

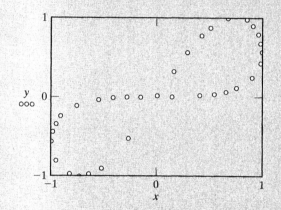

Generalized Linear Regression

The linfit (vx, vy, vf) function performs a linear regression with an arbitrary linear model. For example, we might try to improve the fit of the regression line to the earlier temperature–time data by using a second-order polynomial, such as

$$\text{Temp}_{pred} = b_0 + b_1 \text{ time} + b_2 \text{ time}^2$$

The linfit( ) function will find the coefficients b_0, b_1, and b_2 that best fit the model to the data. However, the linfit( ) function does not handle units, so we first remove the units from the Time and Temp vectors:

$$\text{time} := \frac{\text{Time}}{\text{min}} \qquad \text{temp} := \frac{\text{Temp}}{\text{K}}$$

Then, we define the linear model. The second-order polynomial has three terms: a constant, a term with time raised to the first power, and a term with time raised to the second power. We define this functionality in the vector function f ( ) and then perform the regression by using the linfit( ) function. The coefficients computed by linfit( ) are stored in the vector b. For example, for the time–temperature data, we have

$$f(x) := \begin{bmatrix} 1 \\ x \\ x^2 \end{bmatrix}$$

b := linfit(time, temp, f)

$$b = \begin{bmatrix} 297.721 \\ 1.742 \\ 0.113 \end{bmatrix}$$

We can then use the coefficients with the second-order polynomial to compute predicted temperature values at each time. A range variable, i, was used to compute such a value at each time value:

$$\text{temp}_{\text{pred}_i} := b_0 + b_1 \cdot \text{time}_i + b_2 \cdot (\text{time}_i)^2$$

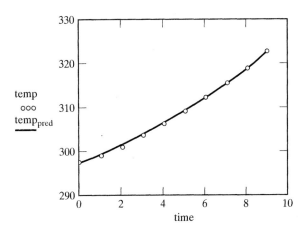

Then the graph was produced by plotting the resulting temperature vector against the associated time vector.

As the preceding graph shows, the second-order polynomial appears to do a much better job of fitting the data than the linear model does. We can quantify the "goodness of fit" using the corr( ) function:

r2 := corr(temp, temp$_{\text{pred}}$)^2
r2 = 0.99961

The R^2 value is closer to 1 than the R^2 we obtained by using simple slope–intercept linear regression, indicating that the polynomial is a better fit than the straight line obtained via simple linear regression.

PRACTICE!

Look at the following data, and decide whether or not to include an intercept in the regression model:

$$X = \begin{bmatrix} 0.0 \\ 0.5 \\ 1.0 \\ 1.5 \\ 2.0 \\ 2.5 \\ 3.0 \end{bmatrix} \quad Y_{\text{clean}} = \begin{bmatrix} 0.00 \\ 0.48 \\ 0.84 \\ 1.00 \\ 0.91 \\ 0.60 \\ 0.14 \end{bmatrix}$$

Note: These are the same data used in the "Practice!" box for the spline fit (to save typing).

Now try using the following polynomial models to fit the data with the `linfit()` function:

a. $y_p = b_0 + b_1 x + b_2 x^2$ or, $y_p = b_0 x + b_1 x^2$

b. $y_p = b_0 + b_1 x + b_2 x^2 + b_3 x^3$ or, $y_p = b_0 x + b_1 x^2 + b_2 x^3$

Other Linear Models

The models used in the earlier examples, namely,

$$\text{temp}_{\text{pred}} = b + m \cdot \text{time}$$

and

$$\text{temp}_{\text{pred}} = b_0 + b_1 \cdot \text{time} + b_2 \cdot \text{time}^2$$

are both linear models (linear in the coefficients, not in time). Since `linfit( )` works with any linear model, you could try fitting an equation such as

$$\text{temp}_{\text{pred}} = b_0 + b_1 \cdot \sinh(\text{time}) + b_2 \cdot \text{atan}(\text{time}^2)$$

or

$$\text{temp}_{\text{pred}} = b_0 \cdot \exp(\text{time}^{0.5}) + b_1 \cdot \ln(\text{time}^3)$$

These equations are linear in the coefficients (the b's) and are compatible with `linfit( )`. The `f(x)` functions for the respective equations would look like this:

$$f(x) := \begin{bmatrix} 1 \\ \sinh(x) \\ \text{atan}(x) \end{bmatrix} \quad f(x) := \begin{bmatrix} \exp(x^{0.5}) \\ x^3 \end{bmatrix}$$

There is no reason to suspect that either of these last two models would be a good fit to the temperature–time data. In general, you choose a linear model either from some theory that suggests a relationship between your variables or from looking at a plot of the data set.

PRACTICE!

A plot of the data used in the spline fit and polynomial regression "Practice!" boxes has a shape something like half a sine wave. Try fitting the data with a linear model such as

a. $y_p = b_0 + b_1 \sin(x)$
b. $y_p = b_0 + b_1 \cos(x)$

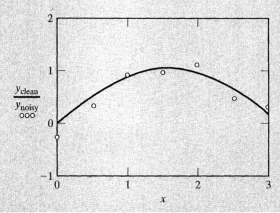

Do the following data suggest that the model should include an intercept?

$$x = \begin{bmatrix} 0.0 \\ 0.5 \\ 1.0 \\ 1.5 \\ 2.0 \\ 2.5 \\ 3.0 \end{bmatrix} \quad y_{\text{clean}} = \begin{bmatrix} 0.00 \\ 0.48 \\ 0.84 \\ 1.00 \\ 0.91 \\ 0.60 \\ 0.14 \end{bmatrix} \quad y_{\text{noisy}} = \begin{bmatrix} -0.25 \\ 0.33 \\ 0.88 \\ 0.92 \\ 1.07 \\ 0.44 \\ 0.25 \end{bmatrix}$$

APPLICATIONS: FITTING PHYSICAL PROPERTY DATA

Silane is an interesting chemical that is being increasingly employed in the manufacture of silicon wafers and chips used in the electronics industry. It is highly flammable and, under the right conditions, will ignite spontaneously upon contact with air. The manufacturers of silane must take special precautions to minimize the risks associated with this gas.

The following table gives the vapor pressure of silane, in pascals, at various temperatures, in kelvins:

SILANE Temp. (K)	VAPOR PRESSURE P_{VAPOR}(Pa)
88.48	21
97.54	167
106.60	640
115.66	2227
124.72	6110
133.79	14144
142.85	30318
151.91	57430
160.97	100078
170.03	163913
179.09	254015
188.15	378009
197.21	541607
206.27	754829
215.33	1027536
224.40	1372259
233.46	1804187
242.52	2341104
251.58	3006269
260.64	3827011
269.70	4838992

Analyzing the risks associated with possible accidents is an important part of a chemical facility's safety program and a big part of some chemical engineers' jobs. There are computer programs to help perform risk analyses, but they require some knowledge of the chemical, physical, and biomedical properties of the chemicals involved. Risk analysis programs have been written to use standardized fitting equations, so if you add your own information, it must be in a standard form. The accompanying silane vapor pressure data were calculated by using the ChemCad Physical Properties Database.[1]

STANDARD FITTING EQUATION FOR VAPOR PRESSURE[2]

The standard equation for fitting a curve to vapor pressure p^* is

$$\ln(p^*) = a + \frac{b}{T} + c\ln(T) + dT^e$$

You must specify a and b; a, b; and c; or $a, b, c, d,$ and e. If e is used the equation becomes nonlinear, and other techniques are used to estimate the value of e. For silane, the value of e is expected to be 1.

To try fitting these data by using the first three terms ($a, b,$ and c), we would use the linfit( ) function, with the f( ) vector defined as

$$f(x) := \begin{bmatrix} 1 \\ \dfrac{1}{x} \\ \ln(x) \end{bmatrix}$$

[1] ChemCad is a product of Chemstations, Inc., Houston, TX.
[2] *ChemCad User Guide*, version III, p. 11.64.

The coefficients themselves would be computed by linfit( ) by using the *vectorize* operator (from the Matrix Toolbar) to take the natural logarithm of each element of the P_{vapor} vector:

$$\text{coeffs} := \text{linfit}(\text{Temp}, \overrightarrow{\ln(P_{vapor})}, f)$$

$$\text{coeffs} = \begin{bmatrix} 30.303 \\ -1.802 \cdot 10^3 \\ -1.494 \end{bmatrix}$$

Finally, we can plot the original data and the predicted vapor pressures by using the computed coefficients, just to make sure that the process worked and to verify that the last term is not required:

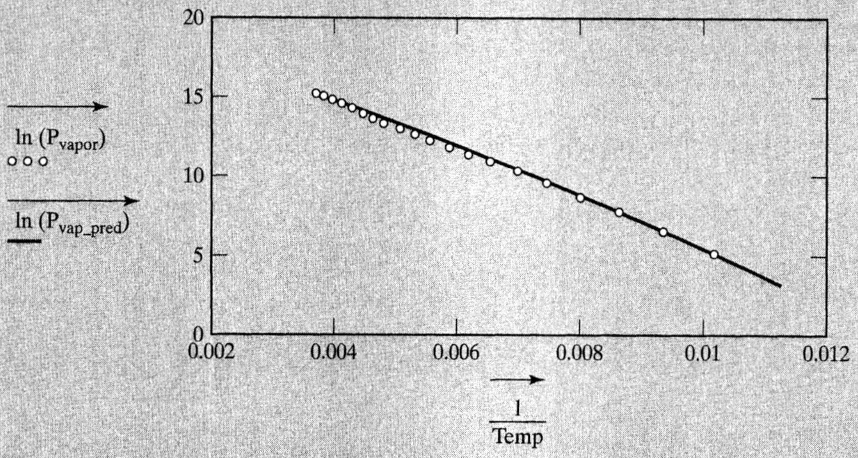

Specialized Regression Equations

Mathcad provides functions for finding coefficients for a number of commonly used fitting equations. These functions use iterative methods to find the coefficients that best fit the data, so a set of initial guesses for the coefficients must be provided.

Note: for more information on interative solutions, see Section 8.1.

As an example, consider fitting the data shown here with an exponential curve:

$$x = \begin{pmatrix} 0 \\ 1 \\ 2 \\ 3 \\ 4 \\ 5 \\ 6 \\ 7 \\ 8 \\ 9 \\ 10 \end{pmatrix} \quad y = \begin{pmatrix} 5.32 \\ 5.83 \\ 6.09 \\ 7.12 \\ 7.62 \\ 9.15 \\ 9.95 \\ 11.8 \\ 13.39 \\ 15.19 \\ 18.12 \end{pmatrix}$$

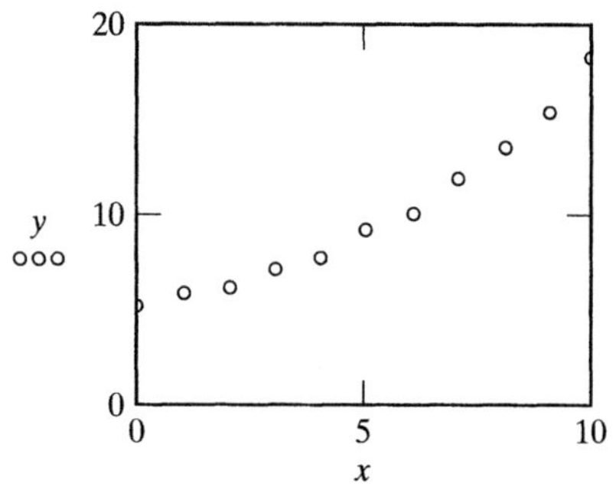

The data have been plotted, and the y values do seem to increase exponentially (albeit, weakly), so an *exponential curve* may be a good choice.

The `expfit( )` function finds values for a, b, and c that best fit the following equation to the data:

$$y_{pred} = a\,e^{bx} + c$$

Note: In older versions of Mathcad (Before Mathcad 11), in order to use `expfit( )` a vector of initial guesses for a, b, and c had to be provided:

$$\text{guesses} := \begin{pmatrix} 1 \\ 1 \\ 1 \end{pmatrix}$$

Then the vector of x values and the vector of y values (and the guesses if desired) are sent to `expfit( )`. The coefficients of the exponential equation are returned:

```
coeffs := expfit (x, y)
```

$$\text{coeffs} = \begin{pmatrix} 2.398 \\ 0.184 \\ 2.854 \end{pmatrix}$$

These coefficients can be used with the known x values in the fitting equation to predict values:

```
a := coeffs₀       a = 2.398
b := coeffs₁       b = 0.184
c := coeffs₂       c = 2.854
Y_pred := a·e^(b·x)+c
```

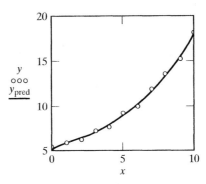

The following table summarizes the specialized fitting equations available in Mathcad:

	FUNCTION	EQUATION OF CURVE
Exponential	expfit(vx, vy, vg)	$y_{pred} = a\,e^{bx} + c$
Logistic	lgsfit(vx, vy, vg)	$y_{pred} = \dfrac{a}{(1 + b e^{-\alpha})}$
Logarithmic	logfit(vx, vy, vg)	$y_{pred} = a\,\ln(x)^{bx} + c$
Power	pwrfit(vx, vy, vg)	$y_{pred} = a\,x^b + c$
Sine	sinfit(vx, vy, vg)	$y_{pred} = a\,\sin(x + b) + c$

Note that the parameters for each function are the same: a vector of x values, a vector of y values, and a vector of initial guesses for the three coefficients (a, b, c). (Guess values are optional with `expfit( )`).

SUMMARY

In this chapter, you learned to use Mathcad's built-in functions for data analysis. You learned to make *x*–*y* graphs from data, to calculate statistical values from data, to interpolate between data set values, and to fit curves (linear models) to data. You also learned about the risk associated with extrapolating data, especially with nonlinear methods like cubic splines.

MATHCAD SUMMARY

X–Y Graphs Vector Against Vector:

1. Select X–Y Plot from the Graph Toolbar.
2. Enter the name of your vector of independent values on the *x*-axis.
3. Enter the name of your vector of dependent values on the *y*-axis.

Element By Element:

1. Declare a range variable starting at 0 and going to last(v), where v is one of your data set vectors.
2. Select X–Y Plot from the Graph Toolbar.
3. Enter the name of your vector of independent values on the *x*-axis—with an index subscript containing the range variable.
4. Enter the name of your vector of dependent values on the *y*-axis—with an index subscript containing the range variable.

Quick Plot:

1. Select X–Y Plot from the Graph Toolbar.
2. Enter your function name on the *y*-axis, with a dummy variable as an argument.
3. Enter the dummy variable on the *x*-axis.
4. Adjust the *x*-axis limits to change the displayed range.

Statistical Functions:

`mean(A)`	Returns the mean (arithmetic average) of the values in A.
`stdev(A)`	Returns the population standard deviation of the values in A.
`var(A)`	Returns the population variance of the values in A.
`stdev(A)`	Returns the sample standard deviation of the values in A.
`var(A)`	Returns the sample variance of the values in A.

Interpolation:

`linterp(vx, vy, x`$_{new}$`)`	Returns the *y* value corresponding to $x = x_{new}$, computed by using linear interpolation on the *x* and *y* data.
`cspline(vx, vy)`	Returns the vector of second derivatives that specify a spline curve that is cubic at the endpoints.
`pspline(vx, vy)`	Returns the vector of second derivatives that specify a spline curve that is parabolic at the endpoints.
`lspline(vx, vy)`	Returns the vector of second derivatives that specify a spline curve that is linear at the endpoints.

`interp(vs, vx, vy, x_new)` — Uses the vector of second derivatives from any of the spline functions previously listed and returns the y value corresponding to $x = x_{\text{new}}$, which is computed by using spline interpolation on the x and y data.

Regression:

`slope(vx, vy)` — Returns the slope of the best fit (minimum total squared error) straight line through the data in vx and vy.

`intercept(vx, vy)` — Returns the intercept of the best fit straight line through the data in vx and vy.

`corr(vx, vy)` — Returns the coefficient of correlation (usually called R) of the best fit straight line through the data in vx and vy. The coefficient of determination, R^2, can be computed from R and is more commonly used.

`linfit(vx, vy, vf)` — Returns the coefficients that best fit the linear model described by vf to the data in vx and vy. The vector vf is a vector of functions you provide that describes the linear model you want to use to fit to the data.

Specialized Fitting Functions:

`expfit(vx, vy, vg)` — Fits an exponential curve ($y_{\text{pred}} = a\, e^{bx} + c$) to the data in the vx and vy vectors. A three-element vector of guessed values for the coefficients (a, b, c) may be provided (optional).

`lgsfit(vx, vy, vg)` — Fits a logistic curve $y_{\text{pred}} = a/(1 + be^{-cx})$ to the data in the vx and vy vectors. A three-element vector of guessed values for the coefficients (a, b, c) must be provided.

`logfit(vx, vy, vg)` — Fits a logarithmic curve ($y_{\text{pred}} = a \ln(x)^b + c$) to the data in the vx and vy vectors. A three-element vector of guessed values for the coefficients (a, b, c) must be provided.

`pwrfit(vx, vy, vg)` — Fits a power curve ($y_{\text{pred}} = ax^b + c$) to the data in the vx and vy vectors. A three-element vector of guessed values for the coefficients (a, b, c) must be provided.

`sinfit(vx, vy, vg)` — Fits an exponential curve ($y_{\text{pred}} = a \sin(x + b) + c$) to the data in the vx and vy vectors. A three-element vector of guessed values for the coefficients (a, b, c) must be provided.

KEY TERMS

coefficient of determination (R^2)
correlation coefficient (R)
exponential curve
intercept
interpolation
linear interpolation
linear regression
logarithmic curve
logistic curve
mean
population
power curve
Quick Plot
representative sample
risk analysis
sine curve
slope
spliner interpolation
standard deviation
variance
visualization
sample
x–y graph

Problems

1. **STATISTICS: COMPETING COLD REMEDIES**

   Two proposed cold remedies—CS1 (cough syrup) and CS2 (chicken soup)—were compared in a hospital study. The objective was to determine which treatment enabled patients to recover more quickly. The duration of sickness was defined as the time (days) between when patients reported to the hospital requesting treatment and when they were discharged from the hospital. The results are tabulated as follows:

CS1		CS2	
Patient	Duration	Patient	Duration
1	8.10	1	5.79
2	7.25	2	2.81
3	5.89	3	3.27
4	7.12	4	4.87
5	5.60	5	5.49
6	1.85	6	6.60
7	4.00	7	2.63
8	5.58	8	4.64
9	9.92	9	6.96
10	4.25	10	5.83

   a. Determine the mean recovery time for each medication.
   b. Which remedy was associated with the fastest recoveries?

2. **STATISTICS: QUALITY CONTROL**

   A beverage bottling plant uses a sampling protocol for quality control. At random intervals, a bottle is pulled off the line and the volume is measured to make sure that the filling operation is working correctly. The company has several criteria it watches:

   - Ideally, the company wants to put an average of 2.03 liters in each bottle. Since the plant must shut down whenever it is found that less than the labeled amount (on average) is being put in each bottle, setting the target amount higher than 2.0 liters reduces the number of times the company

2.00	2.09
1.99	2.08
2.00	2.09
1.97	2.07
1.93	1.97
2.03	2.03
2.02	1.99
2.06	2.07
2.16	2.02
2.08	1.99

has to shut down the process to recalibrate the equipment. If the average fill volume is less than 2.03 liters, but greater than 2.00 liters, the equipment will not be shut down immediately, but will be recalibrated the next time the equipment is down for any reason (cleaning, breakage, etc.)
- If the average fill volume exceeds 2.06 liters, the equipment must be shut down for recalibration to prevent bottles from overflowing.
- If the standard deviation exceeds 40 ml, the equipment must be shut down for recalibration. This is to eliminate customer complaints about incompletely filled bottles, as well as to prevent bottles from overflowing.

Given the sample volumes (liters) in the accompanying table, should the equipment be shut down and recalibrated?

3. USING X–Y TRACE WITH GRAPHS
Create a QuickPlot of the function

$$f(x) = 1 - e^{-x}$$

Use Mathcad's X–Y Trace dialog to evaluate this function at $x = 1, 2$, and 3. (Click on the graph, and then use the menu commands Format/Graph/Trace.)

4. USING QUICKPLOTS TO SOLVE PROBLEMS
The equation describing the process of warming a hot tub by adding hot water is

$$T = T_{IN} - (T_{IN} - T_{START}) e^{-\frac{Q}{V} t},$$

where

T	is the temperature of the water in the hot tub,
T_{IN}	is the temperature of the water flowing into the hot tub, (130°F)
T_{START}	is the initial temperature of the water in the hot tub (65°F),
Q	is the hot-water flow rate (5 gpm),
V	is the volume of the hot tub (500 gallons), and
t	is the elapsed time since the hot water started flowing.

a. Use QuickPlot to see how long it will take the hot tub to reach 110°F. (You may want to work this problem without units, since °F is not defined in Mathcad.)

b. If the hot-water flow rate was increased to 10 gpm, how long would it take for the water temperature in the tub to reach 110°F?

5. **SIMPLE LINEAR REGRESSION**

   Plot each of the following three data sets to see whether a straight line through each set of points seems reasonable:

x	y_1	y_2	y_3
0	2	0.4	10.2
1	5	3.6	4.2
2	8	10.0	12.6
3	11	9.5	11.7
4	14	12.0	28.5
5	17	17.1	42.3
6	20	20.4	73.6
7	23	21.7	112.1

   If a straight line is reasonable, use the `slope( )` and `intercept( )` functions to calculate the regression coefficients for the set. (The same x data have been used in each example to minimize typing.)

6. **CHOOSING A FITTING FUNCTION**

   Half the challenge of regression analysis is choosing the right fitting function. The purpose of this exercise is to demonstrate the general shape of some typical fitting functions and then to have you choose an appropriate function and fit a curve to some data.

   Typical Fitting Functions

   Make QuickPlots of the following functions over the indicated range of the independent variable x:

x	y	FUNCTION	COEFFICIENTS	RANGE
0.4	38.9	$a + bx$	$a = 5$, $b = 1$	$0 < x < 10$
0.9	19.8			
1.4	14.2	$a + bx + cx^2$	$a = 5$, $b = 1$, $c = 1$	$0 < x < 10$
1.9	10.8			
2.4	9.8	$a + bx + cx^2 + dx^3$	$a = -20$, $b = 50$, $c = -10$, $d = 0.6$	$0 < x < 10$
2.9	7.3			
3.4	7.5			
3.9	4.6	$a + b \ln(x)$	$a = 2$, $b = 1$	$0 < x < 1$
4.4	6.6			
4.9	6.5	$a + be^x$	$a = 2$, $b = 1$	$0 < x < 3$
5.4	5.3			
5.9	5.0	$a + be^{-x}$	$a = 2$, $b = 1$	$0 < x < 3$
6.4	3.3			
6.9	5.3			
7.4	4.9	$a + b/x$	$a = 1$, $b = 1$	$0 < x < 1$

Selecting and Fitting a Function to Data

Now graph the data shown at the left, and select an appropriate fitting function. Use linfit( ) to regress the data with your selected function. Plot the original data and the values predicted by your fitting function on the same graph to visually verify the fit.

7. **FITTING AN EXPONENTIAL CURVE TO DATA**

    In Problem 4, the exponential function describing the process of heating a hot tub was given. If you have experimental temperature vs. time data taken as the hot tub warms up, Mathcad's expfit( ) function could be used to fit an exponential function to the experimental data. The general form of an exponential curve is

    $$y_{pred} = a \cdot e^{b \cdot x} + c$$

    Suppose the hot-tub warming data are as follows:

TIME (minutes)	WATER TEMPERATURE (°F)
0	65
15	75
30	82
45	89
60	95
75	99
90	102

    a. Let

    $$\text{guesses} := \begin{bmatrix} 1 \\ -0.01 \\ 10 \end{bmatrix}$$

    be the initial estimates of the exponential function coefficients. Use expfit( ) to find the coefficients of the exponential curve that fits the experimental data.

    b. Plot the experimental values and the exponential curve on the same graph. Does the curve fit the data?

8. **CHECKING THERMOCOUPLES**

    Two thermocouples have the same color codes on the wires, and they appear to be identical. But when they were calibrated, they were found to give somewhat different output voltages. The calibration data are shown in the following table:

WATER TEMP. (°C)	TC 1 OUTPUT (mV)	TC 2 OUTPUT (mV)
20.0	1.019	1.098
30.0	1.537	1.920
40.0	2.059	2.526
50.0	2.585	2.816
60.0	3.116	2.842
70.0	3.650	4.129
80.0	4.187	4.266
90.0	4.726	4.340

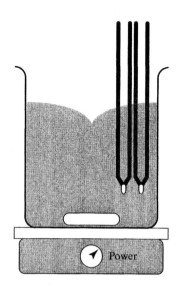

T(°C)	V(mV)
10	0.397
20	0.796
30	1.204
40	1.612
50	2.023
60	2.436
70	2.851
80	3.267
90	3.662

Plot the output voltages (y-axis) against temperature (x-axis) for each thermocouple. Which thermocouple would you want to use in an experiment? Why?

9. THERMOCOUPLE CALIBRATION

Thermocouples are made by joining two dissimilar metal wires. Contact between the two metals results in a small, but measurable, voltage drop across the junction. This voltage drop changes as the temperature of the junction changes; thus, the thermocouple can be used to measure temperature if you know the relationship between temperature and voltage. Equations for common types of thermocouples are available, or you can simply take a few data points and prepare a calibration curve. This is especially easy for thermocouples because, for small temperature ranges, the relationship between temperature and voltage is nearly linear.

a. Use the `slope( )` and `intercept( )` functions to find the coefficients of a straight line through the data at the left.
Note: The thermocouple voltage changes because the temperature changes—that is, the voltage *depends* on the temperature. For regression, the independent variable (temperature) should always be on the x-axis, and the dependent variable (voltage) should be on the y-axis.

b. Calculate a predicted voltage at each temperature, and plot the original data and the predicted values together to make sure that your calibration curve actually fits the data.

10. ORIFICE METER CALIBRATION

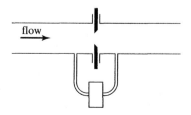

Q FT³/MIN	DP PSI
3.9	0.13
7.9	0.52
11.8	1.18
15.7	2.09
19.6	3.27
23.6	4.71
27.5	6.41
31.4	8.37
35.3	10.59
39.3	13.08

Orifice meters are commonly used to measure flow rates, but are highly nonlinear devices. Because of this nonlinearity, special care must be taken when one prepares calibration curves for these meters. The equation relating the flow rate Q to the pressure drop across the orifice ΔP (the measured variable) is

$$Q = \frac{A_0 C_0}{\sqrt{1 - \beta^4}} \sqrt{\frac{2 g_c \Delta P}{\rho}}$$

For purposes of creating a calibration curve, the details of the equation are unimportant (as long as the other terms stay constant). What is necessary is that we see the theoretical relationship between Q and ΔP, namely,

$$Q \propto \sqrt{\Delta P}$$

Also, the pressure drop across the orifice plate depends on the flow rate, not the other way around. So $\sqrt{\Delta P}$ should be regressed as the dependent variable (y values) and Q as the independent variable (x values).

a. Using the accompanying data, regress Q and $\sqrt{\Delta P}$ to create a calibration curve for the orifice meter.

b. Calculate and plot predicted values together with the original data to check your results.

Note that if you use the slope( ) and intercept( ) functions in this problem, you will need to use the *vectorize* operator as well, because taking the square root of an entire matrix is not defined. The vectorize operator tells Mathcad to perform the calculations (taking the square root, for example) on each element of the matrix. The slope( ) function might then look like this:

$$m := \text{slope}\left(Q, \overrightarrow{\sqrt{\Delta P}}\right)$$

11. **VAPOR–LIQUID EQUILIBRIUM**

When a liquid mixture is boiled, the vapor that leaves the vessel is enriched in the more volatile component of the mixture. The vapor and liquid in a boiling vessel are in equilibrium, and vapor–liquid equilibrium (VLE) data are available for many mixtures. The data are usually presented in tabular form, as in the accompanying table, which represents VLE data for mixtures of methanol and n-butanol boiling at 12 atm. From the VLE data, we see that if a 50:50 liquid mixture of the alcohols is boiled, the vapor will contain about 80% methanol.

VLE data are commonly used in designing distillation columns, but an equation relating vapor mass fraction to liquid mass fraction is a lot handier than tabulated values.

Use the linfit( ) function and the tabulated VLE data to obtain an equation relating the mass fraction of methanol in the vapor (y) to the mass fraction of vapor in the liquid (x). Test different linear models (e.g., polynomials) to see which gives the best fit to the experimental data. For your best

x	y
0.000	0.000
0.022	0.066
0.046	0.133
0.071	0.201
0.098	0.269
0.126	0.335
0.156	0.399
0.189	0.461
0.224	0.519
0.261	0.574
0.302	0.626
0.346	0.674
0.393	0.720
0.445	0.762
0.502	0.802
0.565	0.839
0.634	0.874
0.710	0.907
0.796	0.939
0.891	0.970
1.000	1.000

model, calculate predicted values using the coefficients returned by linfit( ), and plot the predicted values and the original data values on the same graph. Your result should look something like this:

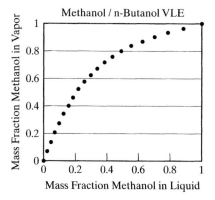

Note: The VLE data shown were generated from version 4.0 of Chemcad.[1] The x column represents the mass fraction of methanol in the boiling mixture. (The mass fraction of n-butanol in the liquid is calculated as 1 − x for any mixture.) The mass fraction of methanol in the vapor leaving the solution is shown in the y column.

12. FITTING PHYSICAL PROPERTY DATA

In an earlier Applications example, experimental data on the vapor pressure of silane were fit to a standard equation. Try fitting a curve to data for silane's liquid heat capacity. The standard fitting equation for this property is

$$C_P = a + bT + cT^2 + dT^3 + eT^4$$

Not all coefficients need to be used (i.e., any order of polynomial is acceptable).

SILANE	
$T_{min}(K)$	88.48
$T_{max}(K)$	161.00

TEMP	C_P	TEMP	C_P
88.48	59874	124.74	61026
92.11	60174	128.37	61231
95.73	60676	131.99	61289
99.36	60156	135.62	61655
102.98	60358	139.24	61160
106.61	60863	142.87	61658
110.24	60604	146.50	62038
113.86	60784	150.12	62085
117.49	61037	153.75	61864
121.11	61215	157.37	61712

---

[1]Chemcad is a chemical process simulation package produced by Chemstations, Inc. in Houston, Texas, USA.

Note: The data set for the liquid heat capacity (J/kmol K) of silane is shown. Find the coefficients for the best fitting model for the data.

13. **INTERPOLATION**

    Use linear interpolation with the thermocouple calibration data in Problem 8 to predict

    a. the thermocouple voltage at a temperature of 85°C.

    b. the temperature of the junction when the thermocouple voltage is 2.500 mV.

    Use a cubic spline interpolation with the orifice meter calibration data in Problem 9 to predict

    c. the flow rate corresponding to a pressure drop of 9.5 psi.

    d. the pressure drop to be expected at a flow rate of 15 ft^3/min.

14. **INTERPOLATING TABULATED DATA**

    In the past, a lot of useful information was provided in the form of tables and graphs. Much of this information is now available in equation form (see the note at the end of this problem), but occasionally it is still necessary to read values from tables. Consider the following *compound amount factor* table:

				COMPOUND AMOUNT FACTORS					
Year	2%	4%	6%	8%	10%	12%	14%	16%	18%
0	1.000	1.000	1.000	1.000	1.000	1.000	1.000	1.000	1.000
2	1.040	1.082	1.124	1.166	1.210	1.254	1.300	1.346	1.392
4	1.082	1.170	1.262	1.360	1.464	1.574	1.689	1.811	1.939
6	1.126	1.265	1.419	1.587	1.772	1.974	2.195	2.436	2.700
8	1.172	1.369	1.594	1.851	2.144	2.476	2.853	3.278	3.759
10	1.219	1.480	1.791	2.159	2.594	3.106	3.707	4.411	5.234
12	1.268	1.601	2.012	2.518	3.138	3.896	4.818	5.936	7.288
14	1.319	1.732	2.261	2.937	3.797	4.887	6.261	7.988	10.147
16	1.373	1.873	2.540	3.426	4.595	6.130	8.137	10.748	14.129
18	1.428	2.026	2.854	3.996	5.560	7.690	10.575	14.463	19.673
20	1.486	2.191	3.207	4.661	6.727	9.646	13.743	19.461	27.393
22	1.546	2.370	3.604	5.437	8.140	12.100	17.861	26.186	38.142
24	1.608	2.563	4.049	6.341	9.850	15.179	23.212	35.236	53.109

    The table can be used to determine the future value of an amount deposited at a given interest rate. One thousand dollars invested at 10% (annual percentage rate) for 20 years would be worth 6.727 × $1,000 = $6,727 at the end of the 20th year. The 6.727 is the compound amount factor for an investment held 20 years at 10% interest.

    Interest tables such as the preceding provided useful information, but it was frequently necessary to interpolate to find the needed value.

    a. Use linear interpolation to find the compound amount factor for 6% interest in year 15.

b. Use linear interpolation to find the compound amount factor for 11% interest in year 10.

Note: Mathcad provides the `fv()` function, that can be used to calculate future values directly, eliminating the need for an interest table. The `fv(rate,N,pmt,pv)` function finds the future value of an amount deposited at time zero (`pv`, include a negative sign to indicate out-of-pocket expenses), plus any periodic payments (`pmt`), invested at a specified interest `rate`, for N periods. The example that was worked out earlier in this problem looks like this in Mathcad:

```
F := fv(10%, 20, 0, -1000)
F = 6727
```

15. **RELATING HEIGHT TO MASS FOR A SOLIDS STORAGE TANK**

Solids storage tanks are sometimes mounted on *load cells* (large scales, basically) so that the mass of solids in the tank is known, rather than the height of the solids. To make sure they do not overfill the tank, the operators might ask for a way to calculate the height of material in the tank, given the mass reading from the load cells.

h(ft)	m(lb)
0	0
2	56
4	447
6	1508
8	3574
10	6981
12	11470
14	16000
16	20520
18	25040
20	29570
22	34090
24	38620

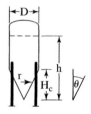

The values shown in the accompanying table relate height and mass in the tank and are the starting point for this problem. The tank has a conical base section ($\theta = 30°$) and a diameter of 12 feet. The apparent density of the solids in the tank is 20 lb/ft^3.

Use a cubic spline to fit a curve to the mass (as `x`) and height (as `y`) data, and then use the `interp( )` function to predict the height of solids in the tank when the load cells indicate 3200 kg solids.

Note: The load cells actually measure the mass of the tank and the stored solids, but adjusting the display to read 0 kg before adding any solids effectively causes the load cells to display only the mass of the solids. This is called setting the *tare weight* for the load cells.

6
Programming in Mathcad

NEW ENERGY SOURCES

The world population continues to burn petroleum products at increasing rates, and at some point the oil reserves will run out. Among the alternative fuels are coal and nuclear fission, as well as largely untapped resources such as solar, wind, and tidal power. Although there are substantial obstacles to overcome if some of these sources are to be effectively utilized, there also are significant problems to solve if we continue to rely on predominantly fossil fuels. The world's energy supply is a major challenge facing the next generation of engineers.

SECTIONS

6.1 Mathcad Programs
6.2 Writing a Simple Program
6.3 The Programming Toolbar
6.4 Program Flowcharts
6.5 Basic Elements of Programming

OBJECTIVES

After reading this chapter, you will

- know what Mathcad programs are, and when they are useful
- know how to write a program in Mathcad
- understand how to access program keywords through the Programming Toolbar
- be able to read and write programming flowcharts
- understand the seven basic elements of all programming languages, and understand how each is implemented in Mathcad

6.1 MATHCAD PROGRAMS

A Mathcad *program region* is essentially a *multistep function*. Program regions are not complicated, but they can be very handy to make complex functions easier to write, and easier for others to comprehend. For example, here is a `thermostat(T)` function that sets a `heater` variable to 1 (on) if the temperature is below 23°C, or −1 (off) if the temperature is above 25°C. If the temperature is between 23 and 25°C, then `heater` is set to 0, implying that no action should be taken.

First, the function is written *without* using a program:

```
thermostat(T) := if(T < 23, 1, if(T > 25, -1, 0))
heater := thermostat(22)
heater = 1

heater := thermostat(26)
heater = -1

heater := thermostat(24)
heater = 0
```

Now, we rewrite the function using a Mathcad program region (to be explained later):

$$\text{thermostat}(T) := \begin{vmatrix} RV \leftarrow 0 \\ RV \leftarrow 1 \text{ if } T < 23 \\ RV \leftarrow -1 \text{ if } T > 25 \end{vmatrix}$$

```
heater := thermostat(22)
heater = 1

heater := thermostat(26)
heater = -1

heater := thermostat(24)
heater = 0
```

Note: The `RV` in the `thermostat` program stands for *return value*. All functions have return values, and the multistep functions that Mathcad calls programs also have return values. By default, whatever is on the last assignment line of the program is the returned value. In the `thermostat` program, that will always be variable `RV`. You can return a different result by using the `return` program statement, which will be described later.

In both cases, the `heater` variable is assigned the same values, but most people would find

$$\text{thermostat}(T) := \begin{vmatrix} RV \leftarrow 0 \\ RV \leftarrow 1 \text{ if } T < 23 \\ RV \leftarrow -1 \text{ if } T > 25 \end{vmatrix}$$

easier to understand than

```
thermostat(T) := if(T < 23, 1, if(T > 25, -1, 0))
```

The following version might be even easier for someone to understand:

$$T_{\text{Cold}} := 23$$
$$T_{\text{Hot}} := 25$$

$$\text{thermostat}(T) := \begin{vmatrix} RV \leftarrow \text{"unchanged"} \\ RV \leftarrow \text{"on"} \text{ if } T < T_{\text{Cold}} \\ RV \leftarrow \text{"off"} \text{ if } T > T_{\text{Hot}} \end{vmatrix}$$

```
heater := thermostat(22)
heater = "on"
```

6.2 WRITING A SIMPLE PROGRAM

Much of this chapter will be spent presenting the basic elements of programming, but first we will develop a very simple program, just to provide an overview of writing a program in Mathcad. Our program will calculate the area of a circle, given the diameter.

Since Mathcad programs are essentially multistep functions, we begin writing a program by defining a function. This includes three pieces:

- function name (which will become the program name)
- parameter (argument) list
- assignment (define as equal to) operator

The *program name* will be used to refer to the program whenever it is needed in the rest of the worksheet. Here, the program is called A_{circle}.

The *parameter list* (or, *argument list*) is the list of all the variable information that must be known before the program can do its job. For example, in order to solve for the area of a circle, we need to know the diameter (and the value of π, but π is a predefined variable in Mathcad):

$$A_{circle}(\texttt{diameter}) := \blacksquare$$

To create a program, simply add two or more lines in the placeholder on the right side of the assignment operator. To do this, use the Add Line button on the Programming toolbar, or the [ ]] key (i.e., press the right-square-bracket key.) In the following example, two placeholders were added by clicking Add Line on the Programming toolbar:

$$A_{circle}(\texttt{diameter}) := \begin{vmatrix} \blacksquare \\ \blacksquare \end{vmatrix}$$

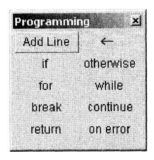

Next, click on the top placeholder and add the formula for calculating a radius from a diameter:

$$A_{circle}(\texttt{diameter}) := \begin{vmatrix} \texttt{radius} \leftarrow \dfrac{\texttt{diameter}}{2} \\ \blacksquare \end{vmatrix}$$

We have created a new variable, `radius`, and assigned it a value. The left-arrow symbol, $\leftarrow$, is the *local definition* operator, which is only used in a program region. It is available either from the Programming toolbar or by pressing the left-brace key [{].

Note: The "local" in local definition operator means that the variable that is being defined and assigned a value (radius in this example) will only exist in the program region. It can be used on any line in the program region, but the rest of the worksheet will not know that the local variable has been created, and will not be able to use it. All *local variables* (variables defined within a program region) disappear once the program stops running.

Finally, use the radius to calculate the area of the circle:

$$A_{circle}(\text{diameter}) := \left| \begin{array}{l} \text{radius} \leftarrow \dfrac{\text{diameter}}{2} \\ \text{area} \leftarrow \pi \cdot \text{radius}^2 \end{array} \right.$$

By default, the value calculated (or set) on the last line of the program (area in this example) is the value returned to the worksheet. So, when the A_{circle} function is used (i.e., when the program is run), as in

```
D := 3·cm
A := A_circle(D)
A = 7.069cm²
```

the value of variable D (with units) is passed into the program through the A_{circle} parameter list. Then the local variable, radius, is calculated (program line 1), and used (line 2) to calculate the area that is passed back out of the program and assigned to the worksheet's variable A.

6.3 THE PROGRAMMING TOOLBAR

The *Programming Toolbar* provides access to the keywords used in Mathcad programs. The toolbar contains the following options (with shortcut keys shown next to each item):

- Add Line [ ]]
- while [CTRL + }]
- otherwise [CTRL + }]
- continue [CTRL + [ ]
- local definition, ← [{]
- for [CTRL + "]
- return [CTRL + |]
- if [}]
- break [CTRL + {]
- on error [CTRL + ']

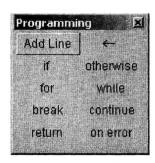

Most of these options will be discussed in subsequent sections as the basic elements of programming are presented. However, the Add Line button is unique to Mathcad, and presented here.

6.3.1 Add Line

The *Add Line button* is used to create placeholders for each of the lines in a program. You can create as many lines as you will need before filling in any of the placeholders, or you can insert or append lines as needed. To insert a placeholder for a new line ahead of an existing line, select the existing line and get the insertion line (vertical blue bar) at the left side:

$$\text{MyProgram}(a,b,c) := \begin{vmatrix} \text{localX} \leftarrow a^2 \\ \text{localY} \leftarrow a + b \\ \text{localX} + \text{localY} \end{vmatrix}$$

Then, click Add Line:

$$\text{MyProgram}(a,b,c) := \begin{vmatrix} \text{localX} \leftarrow a^2 \\ \text{localY} \leftarrow a + b \\ \blacksquare \\ \text{localX} + \text{localY} \end{vmatrix}$$

To add a line to the end of the program, get the insertion line at the right side of the last program line:

$$\text{MyProgram}(a,b,c) := \begin{vmatrix} \text{localX} \leftarrow a^2 \\ \text{localY} \leftarrow a + b \\ \text{localX} + \text{localY} \end{vmatrix}$$

Then, click Add Line:

$$\text{MyProgram}(a,b,c) := \begin{vmatrix} \text{localX} \leftarrow a^2 \\ \text{localY} \leftarrow a + b \\ \text{localX} + \text{localY} \\ \blacksquare \end{vmatrix}$$

6.4 PROGRAM FLOWCHARTS

A *flowchart* is a visual depiction of a program's operation. It is designed to show, step by step, what a program does. Typically, a flowchart is created before the program is written. The flowchart is used by the programmer to assist in developing the program and by others to help understand how the program works.

Among the standard symbols used in computer flowcharts are the following:

SYMBOL	NAME	USAGE
⬭	**Terminator**	Indicates the start or end of a program.
▭	**Operation**	Indicates a computation step.
▱	**Data**	Indicates an input or output step.
◇	**Decision**	Indicates a decision point in a program.
○	**Connector**	Indicates that the flowchart continues in another location.

These symbols are connected by arrows to indicate how the steps are connected, and the order in which the steps occur.

Flowcharting Example

As an example of a simple flowchart, consider the thermostat program:

$$\text{thermostat}(T) := \begin{vmatrix} RV \leftarrow 0 \\ RV \leftarrow 1 \text{ if } T < 23 \\ RV \leftarrow -1 \text{ if } T > 25 \end{vmatrix}$$

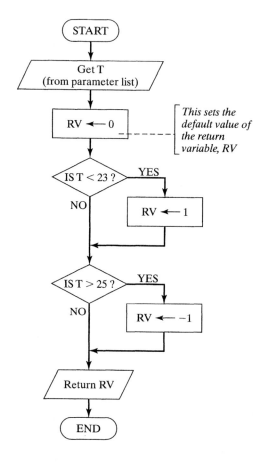

This program's flowchart is shown above. It indicates that the program starts (when Mathcad encounters the program region while evaluating the worksheet), and the value of T is passed into the program from the parameter list. Then, the first line of the program is an operation step: RV ← 0. This step ensures that the return variable, RV, has a value no matter what T value is passed into the program. This is called assigning a *default value* to RV.

The next line in the program is a decision step, indicated by the diamond symbol on the flowchart. The if T < 23 portion of the second program line is the condition that is checked, and the left side of the line, RV ← 1 is the operation that is performed if the condition is found to be true.

The last line of the program is another decision step. The condition if T > 25 is tested, and if the condition is found to be true, the operation RV ← −1 is performed.

Just before the program ends, the value of RV is returned through `thermostat` so that it is available to the worksheet.

You might have noticed that the `thermostat` program tests T twice. Even if it has already been determined that T < 23, it still checks to see if T > 25. This is not a particularly efficient way to write a program. The last steps of a better version might look like this:

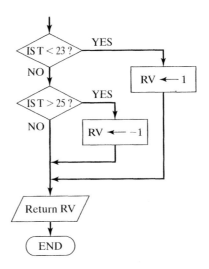

This illustrates two of the reasons for using flowcharts:

- to help identify inefficient programming;
- to indicate how the program should be written.

PRACTICE!

Flowcharts may be used to depict virtually any multistep process, and they are frequently used to illustrate a *decision tree*, or process leading to a particular decision. This Practice! exercise is about creating a flowchart to show a common decision process: determining if someone has a fever and should take some medicine, and if that medicine should be aspirin.

Disclaimer: There is no universally accepted criterion for deciding whether someone's body temperature is high enough to require medication. The values given here are sometimes used, but certainly are no replacement for good medical advice:

- For babies less than a year old, if their temperature is over 38.3°C (101°F), they should receive medication.
- For children less than 12 years old, if their temperature is over 38.9°C (102°F), they should receive medication.
- For people over 12 years old, if their temperature is over 38.3°C (101°F), they should receive medication.

Oral or ear temperatures are assumed in all cases. The threshold is slightly higher for children because their body temperatures fluctuate more than baby or adult temperatures.

Aspirin is a good fever reducer, but you should not give it to children under 12 years old (some say under 19 years) because of the risk of a rare but serious illness known as Reye's syndrome. Acetaminophen and ibuprofen are alternatives for young people.

> Create a flowchart that illustrates the input values, and the decision steps necessary to determine
>
> a. whether the person's fever is high enough to warrant medication, and
> b. whether the medication should be aspirin.
>
> How would your flowchart need to be modified to include a check to see if the person's temperature is above 41.1°C (106°F) since this level requires immediate medical assistance?

6.5 BASIC ELEMENTS OF PROGRAMMING

There are a few elements that are common to all programming languages:

- *Data* — Single-valued variables and array variables are used to hold data.
- *Input* — Getting information into the program is an essential first step in most cases.
- *Operations* — These may be as simple as addition and subtraction, but operations are essential elements of programming.
- *Output* — Once you have a result, you need to do something with it. This usually means assigning the result to a variable, saving it to a file, or displaying the result on the screen. Only the first option is available in Mathcad programs.
- *Conditional Execution* — The ability to have a program decide how to respond to a situation is a very important and powerful aspect of programming.
- *Loops* — Loop structures make repetitive calculations easy to perform in a program.
- *Functions* — The ability to create reusable code elements that (typically) perform a single task is considered an integral part of a modern programming language.

Mathcad provides each of these elements. Access to many of these elements is provided by the *Programming Toolbar*.

6.5.1 Data

What a program does is manipulate data, stored in variables. The data values can

- come from variables on the worksheet,
- come from parameter values passed into the program through the program's parameter list (argument list),
- be read from an external file (covered in the Input section),
- be computed by calculations within the program.

The variables can hold a single value or an array of values.

Using Worksheet Variables in Programs

Mathcad programs work with variables in nearly the same way they are used in a Mathcad worksheet. In fact, variables assigned values on the worksheet may be used in your program, as long as the variables are assigned values before the program is run. In a typical Mathcad worksheet, this means that the variables must be defined above the program.

Note: Just because you can use worksheet variables in a Mathcad program does not mean it is necessarily a good idea. One of the reasons for writing programs is to be able to copy and use the program other places (i.e., in other worksheets). If your program depends on certain variables being predefined in the worksheet before the program runs, then every worksheet in which the program is used would have to include the same predefined variables. A preferred approach is to pass all of the information required by the program into the program via the parameter list. This will be discussed further in the next section.

Variables defined within a program can be used only within the program. In programming, variables that are only defined within a program unit are called *local* variables. The *local definition symbol* in a Mathcad program is a left arrow, ←. It is available either from the programming toolbar, or by typing [{] (left brace key).

Consider the following example, in which a variable A is defined above a program region and is used within the program (C is both the program name and the variable that receives the program's return value):

$$A := 12 \cdot cm^2$$

$$C := \begin{vmatrix} B \leftarrow A + 2 \cdot cm^2 \\ B \end{vmatrix}$$

$$B = \blacksquare \; cm^2$$

The variable B is defined within the program (on the right side of the vertical line). Inside the program it has a value, but if you try to type [B] [=] on the worksheet to display the value of B after the program has run, you will get an error because B has not been defined on the worksheet (only inside the program). However, the B on the last line of the program causes the value of B to be returned to the variable named on the left side of the program definition, C.

You also need to be aware that Mathcad keeps variables defined on the worksheet separate from variables defined inside programs. In the following example, A is defined on the worksheet *and* in the program (since Mathcad keeps them separate, this is allowed):

$$A := 12 \cdot cm^2$$

$$C := \begin{vmatrix} A \leftarrow A + 2 \cdot cm^2 \\ A \end{vmatrix}$$

$$C = 14 \, cm^2$$

The first line of the program in the previous example actually uses both of the A variables. The worksheet's A is used on the right as part of the calculation that is used in the definition of the program's A (on the left of the ←).

To summarize:

- Variables defined on the worksheet above the program region may be used in the program, but their values will not be changed by the program.
- Variables defined within a program can only be used within that program and will disappear when the program ends. (However, their value may be returned from the program.)

Passing Values through a Parameter List

It is good programming practice to pass all of the information required by a function into the function through the parameter list. This allows the function to be self-contained and ready to be used anywhere. This wisdom also applies to Mathcad programs. If they are self-contained, they can be copied and reused wherever they are needed.

While you are writing a program, the variables you include in a parameter list and then use in the program definition are simply there to indicate how the various parameter values should be manipulated by the program. That is, while you are defining the program, the variables in the parameter list are dummy variables. You can use any variables you want, but well-named variables will make your program easier to understand.

The following two program definitions are functionally equivalent, although the first one is preferred because the variable names have more meaning:

```
CylinderAreaV1(D,L) :=  | R ← D/2
                        | A_circle ← π·R²
                        | A_side ← π·D·L
                        | A_surface ← 2·A_circle + A_side
```

$\text{CylinderAreaV1}(3 \cdot \text{cm}, \ 7 \cdot \text{cm}) = 80.111 \text{cm}^2$

```
CylinderAreaV2(v1,v2) :=  | v3 ← v1/2
                          | v4 ← π·v3²
                          | v5 ← π·v1·v2
                          | v6 ← 2·v4 + v5
```

$\text{CylinderAreaV2}(3 \cdot \text{cm}, \ 7 \cdot \text{cm}) = 80.111 \text{cm}^2$

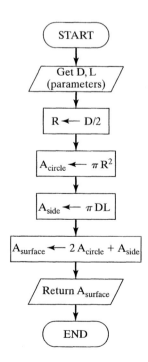

Also, if the variables used in the program definition already have their own definitions, it doesn't matter. While the program is being defined, Mathcad is not calculating anything, just defining the program. In the following example, D and L are defined before the program is defined (that doesn't hurt anything, but the values of D and L are not used while the program is being defined):

$$D := 3 \cdot cm \qquad L := 7 \cdot cm$$

$$CylinderAreaV1(D, L) := \begin{vmatrix} R \leftarrow \dfrac{D}{2} \\ A_{circle} \leftarrow \pi \cdot R^2 \\ A_{side} \leftarrow \pi \cdot D \cdot L \\ A_{surface} \leftarrow 2 \cdot A_{circle} + A_{side} \end{vmatrix}$$

Any defined values of D and L would be used when the program is run (i.e., when the multistep function CylinderAreaV1 ( ) is used):

$$D := 3 \cdot cm \qquad L := 7 \cdot cm$$
$$CylinderAreaV1(D, L) = 80.111 cm^2$$

When you include variables in a parameter list, only the *values* assigned to those variables are actually passed into the program. In programming terms, this is called *passing arguments by value*. The standard programming alternative is *passing arguments by address*, but this is not available in Mathcad. What this means is that you cannot permanently change the value of a variable passed through the parameter list. You can change its value within the program, but once the program has completed, the passed variable will still retain its original value.

Consider this sample program, which passes in the value of variable L, attempts to change its value to 22 watts, and returns the changed value:

$$L := 7 \cdot cm$$

$$oddProgram(L) := \begin{vmatrix} L \leftarrow 22 \cdot watt \\ L \end{vmatrix}$$

$$oddProgram(L) = 22W$$
$$L = 7cm$$

The program does return a value of 22 watts, but the worksheet's value of L = 7 cm is unchanged.

This is what happened: The value of variable L was passed into the program (to the right side of the vertical bar). The program was not given access to the actual variable stored on the worksheet; only the value 7 cm was passed into the program. Then, inside the program, a temporary definition (←) was used. This created a new variable, also called L, but known only within the program. This new variable was assigned a value of 22 watts. The value of 7 cm that was passed into the program was never actually used at all.

In sum, passing data into a program by means of a parameter list

- causes the values of the variables in the parameter list, not the variables themselves, to be passed into the program. This means that a Mathcad program cannot change any value on the worksheet, it can only return a result.
- is generally the preferred way to pass data from the worksheet into a program.
- helps make the programs self-contained so that they can be more easily reused in other worksheets.

APPLICATION: CALCULATING A RESULTANT FORCE AND ANGLE

Force balance problems often require resolving multiple force vectors into horizontal and vertical force components. The solution may then be obtained by summing the force components in each direction and solving for the resultant force and angle. This multistep solution process can be written as a Mathcad program.

The resultant(vh, vv) program will pass two vectors into the program: vh, a vector of horizontal force components; and vv, a vector of corresponding vertical force components. The solution process, as shown by the flowchart, requires that the components in each direction be summed, that the magnitude of the resultant force be computed using the Pythagorean theorem, and that the angle of the resultant force be calculated using the atan() function:

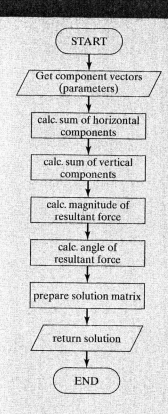

$$vh := \begin{pmatrix} 120 \\ -15 \\ 35 \\ -100 \\ 67 \end{pmatrix} \qquad vv := \begin{pmatrix} 60 \\ -35 \\ -20 \\ 75 \\ 54 \end{pmatrix}$$

$$\text{resultant}(vh, vv) := \begin{vmatrix} \text{sumvh} \leftarrow \sum vh \\ \text{sumvv} \leftarrow \sum vv \\ RF \leftarrow \sqrt{\text{sumvh}^2 + \text{sumvv}^2} \\ \theta \leftarrow \text{atan}\left(\dfrac{\text{sumvv}}{\text{sumvh}}\right) \\ \begin{pmatrix} RF \\ \theta \end{pmatrix} \end{vmatrix}$$

Notes:

1. This problem is being worked without units because two results with different units are being returned. Typically, Mathcad programs can handle units, but when multiple results are returned as a vector, each result must have the same units.

2. The summation operators used in the first two lines of the program are Mathcad's vector sum operators from the Matrix toolbar. These operators provide a quick way to calculate the sum of the values in a vector.

When the resultant program is run, the magnitude and direction of the resultant vector are returned. The angle is in radians (Mathcad's default), but here it has been converted to degrees using the deg unit:

$$\begin{pmatrix} RF \\ \theta \end{pmatrix} := \text{resultant}(vh, vv)$$

$RF = 171.479$
$\theta = 51.392 \text{ deg}$

6.5.2 Input

There are a variety of *input sources* available on computers: keyboard, disk drives, tape drive, a mouse, microphone, and more. Because Mathcad programs are housed within a worksheet, there are basically two available input sources: data available on the worksheet itself, and data in files. The use of worksheet data was covered in the previous section, so this section deals only with reading data files.

Mathcad provides a file input component for importing data to the worksheet, but if you want to read data directly into a Mathcad program, you will need to use the READPRN() function. This function works with text files, and is designed to read an array of values.

Creating a Test Data File

Before we can read a file into a program, we must have a text file containing some data. Consider the following array, created on the Mathcad worksheet:

$$A := \begin{pmatrix} 1 & 5 & 10.12345678 \\ 2 & 6 & 11 \\ 3 & 7 & 12 \end{pmatrix}$$

This data can be saved to a file called TestData.txt on the A: drive, using the WRITEPRN() function from the worksheet:

$$\text{WRITEPRN}(\text{"A:\textbackslash TestData.txt"}) := A$$

The text file that was created looks like this when opened in a text editor such as the Windows Notepad:

```
1       5       10.12
2       6       11
3       7       12
```

By default, the WRITEPRN() function writes the data with four significant figures (10.12, not 10.12345678) into columns that are eight characters wide. This can be changed using the PRN file settings under Tools/Worksheet Options.

Reading Data into a Program

Once a data file exists, it can be read into a program using the READPRN() function:

$$\text{MyProgram}(\text{filePath}) := \begin{vmatrix} B \leftarrow \text{READPRN}(\text{filePath}) \\ RV \leftarrow B^{(0)} \end{vmatrix}$$

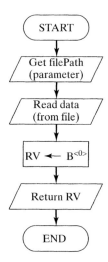

In this program, the file name and drive location are passed into the program through variable `filePath`. The file is read and the data is assigned to local variable B. On the last line of the program the left column of array B is selected using the column operator, and assigned to variable RV to be returned by the program.

When the program is run, the `TestData.txt` file is read, and the left column is returned:

$$\text{FirstColumn} := \text{MyProgram}(\text{"A:}\backslash\text{TestData.txt"})$$

$$\text{FirstColumn} = \begin{pmatrix} 1 \\ 2 \\ 3 \end{pmatrix}$$

Notes:

1. RV was being used in the previous program to indicate where the program's return value is set, but the use of that particular variable name (or any variable at all) is not required. The following program returns exactly the same result:

$$\text{MyProgram}(\text{filePath}) := \begin{vmatrix} B \leftarrow \text{READPRN}(\text{filePath}) \\ B^{\langle 0 \rangle} \end{vmatrix}$$

2. Reading values into a Mathcad program is possible, but the more standard way of using Mathcad is to import the values into the worksheet using a file input component (read component), assign the data to one or more array variables, and then pass the array(s) into the program through the parameter list.

6.5.3 Operations

Most of Mathcad's standard *operators* can be used in programs. These operators are listed in the table that follows. The exceptions are the "define as equal to" operators, := and ≡, which may not be used inside a program. Only the local definition operator, ←, may be used inside of a program.

Standard Math Operators

SYMBOL	NAME	SHORTCUT KEY
+	Addition	+
−	Subtraction	−
*	Multiplication	[Shift-8]
/	Division	/
e^x	Exponential	
1/x	Inverse	
x^y	Raise to a Power	[^], or [Shift-6]
n!	Factorial	!
|x|	Absolute Value	
$\sqrt{}$	Square Root	\
$\sqrt[n]{}$	N^{th} Root	[Ctrl-\]

Matrix Operators

SYMBOL	NAME	SHORTCUT KEY
+	Addition	+
−	Subtraction	−
*	Multiplication	[Shift-8]
/	Division	/
A_n	Array Element	[
M^{-1}	Matrix Inverse	
$|M|$	Determinant	|
→	Vectorize	[Ctrl- −]
$M^{<x>}$	Matrix Column	[Ctrl-6]
M^T	Matrix Transpose	[Ctrl-1]
$M_1 \cdot M_2$	Dot Product	°, or [Shift-8]
$M_1 \times M_2$	Cross Product	[Ctrl-8]

Operator Precedence Rules

Mathcad evaluates expressions from left to right (starting at the assignment operator, :=), following standard *operator precedence rules*.

Operator Precedence

PRECEDENCE	OPERATOR	NAME
First	∧	Exponentiation
Second	*, /	Multiplication, Division
Third	+, −	Addition, Subtraction

For example, you might see the following equation in a Mathcad worksheet:

$$C := A \cdot B + E \cdot F$$

You would need to know that Mathcad multiplies before it adds (operator precedence), in order to understand that the equation would be evaluated as

$$C := (A \cdot B) + (E \cdot F)$$

It is a good idea to include the parentheses to make the order of evaluation clear.

Local Definition Operator, ←

Variables may be defined and assigned values inside a program, but such variables can only be used inside the program (they are of *local scope*) and lose their values when the program terminates (they are *temporary variables*). The *local definition operator* is used to assign values to these local variables. You can either use the Programming toolbar to enter the local definition operator, or press the left-brace key, [{]. The local definition operator is the only way to assign a value to a variable inside a program; the "define as equal to" operator, :=, cannot be used inside a program (except, of course, right after the program name.)

APPLICATION: ALTERNATIVE FUEL CALCULATIONS

According to information provided by the Energy Information Administration,[1] a branch of the US Department of Energy, the people of the world use over 150 quadrillion BTUs (150×10^{15} BTU = 158×10^{15} kJ) of energy from petroleum products each year. To reduce the use of petroleum, some of that energy would need to come from alternative sources such as solar or wind energy. This problem considers what it would take to replace 10% of the current world petroleum usage using today's solar and wind technologies.

SOLAR POWER

A quick survey of available solar panels found some panels (Free Energy Europe, model FEE-20-12) that can produce, at peak light intensity, 54 watts per square meter of panel surface. Assuming that the panels will receive peak light intensity for eight hours per day (a generous assumption), how many square meters of solar panels would be required to replace 10% of the world's petroleum usage?

$$Q_{world} := 150 \cdot 10^{15} \cdot BTU$$
$$Q_{world} = 1.583 \times 10^{17} \text{ kJ}$$
$$q_{solar} := 54 \cdot \frac{watt}{m^2}$$

$\text{solar}(Q_{world}, q_{solar}) :=$

$$\begin{vmatrix} Q_{replace} \leftarrow Q_{world} \cdot 10\% \\ Q_{1\_panel} \leftarrow q_{solar} \cdot 8 \cdot hr \\ A_{panels} \leftarrow \dfrac{Q_{replace}}{Q_{1\_panel}} \\ A_{panels} \end{vmatrix}$$

[1] The Energy Information Administration's website can be found at: <http://www.eia.doe.gov>

$$Area_{required} := \text{solar}(Q_{world}, q_{solar})$$
$$Area_{required} = 10.176 \times 10^{12} \text{ m}^2$$
$$Area_{required} = 10.176 \times 10^{6} \text{ km}^2$$

The required area, using these particular solar panels, is 10 million square kilometers. This is about 15% of the land surface area of the planet.

WIND ENERGY

How many wind 250-kW turbines are required to replace 10% of the world's petroleum usage? Assume favorable winds keep the turbines running 75% of the time. We have

$$kW := 1000 \cdot watt$$
$$Q_{world} := 150 \cdot 10^{15} \cdot BTU$$
$$Q_{world} = 1.583 \times 10^{17} \text{ kJ}$$
$$q_{wind} := 250 \cdot kW.$$

$\text{wind}(Q_{world}, q_{wind}) :=$

$$\begin{vmatrix} Q_{replace} \leftarrow Q_{world} \cdot 10\% \\ Q_{1\_turbine} \leftarrow q_{wind} \cdot 24 \cdot hr \cdot 75\% \\ N_{turbines} \leftarrow \dfrac{Q_{replace}}{Q_{1\_turbine}} \\ N_{turbines} \end{vmatrix}$$

$$N_{required} := \text{wind}(Q_{world}, q_{wind})$$
$$N_{required} = 976.904 \times 10^{6}$$

It would take nearly a billion wind turbines to replace 10% of the world's petroleum usage.

Note: Neither of these examples is intended to suggest that it is futile to try to develop alternative energy sources. They were selected to illustrate that current technologies will require dramatic improvements if they are to succeed. That is, developing alternative energy technologies will provide lots of opportunities for good engineering work in the years to come!

6.5.4 Output

Output from a Mathcad program is very simple—you can only return values from the program back to the worksheet. You cannot assign values to worksheet variables from within a program, and you cannot write values to a file from within a program.

By default, the last value assigned in a program is the program's *return value*. This is usually the assignment on the last line of the program, such as

Section 6.5 Basic Elements of Programming

$$\text{Cylinder Area(D, L)} := \begin{vmatrix} R \leftarrow \dfrac{D}{2} \\ A_{circle} \leftarrow \pi \cdot R^2 \\ A_{side} \leftarrow \pi \cdot D \cdot L \\ A_{surface} \leftarrow 2 \cdot A_{circle} + A_{side} \end{vmatrix}$$

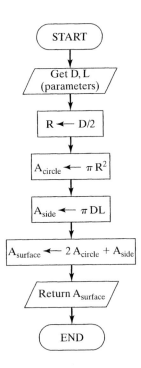

In the `CylinderArea` program, $A_{surface}$ is the last assignment statement, so that is the value returned by the program.

The only time the value of the variable on the last line might not be the return value is if the variable on the last line is never assigned a value. An example of this is the `thermostat` program introduced earlier:

$$\text{thermostat(T)} := \begin{vmatrix} RV \leftarrow 0 \\ RV \leftarrow 1 \text{ if } T < 23 \\ RV \leftarrow -1 \text{ if } T > 25 \end{vmatrix}$$

The assignment of -1 to RV on the last line only happens if $T > 25$. Otherwise, no assignment is made on the last line of the program. A slight modification of this program might make this more apparent:

$$\text{thermostat(T)} := \begin{vmatrix} A \leftarrow 0 \\ B \leftarrow 1 \text{ if } T < 23 \\ C \leftarrow -1 \text{ if } T > 25 \end{vmatrix}$$

Now, C is only assigned a value if $T > 25$. If T is not greater than 25, then this program's return value is held in variable B (if $T < 23$) or A; C is never assigned a value at all.

So, generally, a program's return value is the value calculated or assigned on the last line of the program, unless no value is calculated or assigned on the last line of the program.

Return Statement

You can use a `return` statement to override the default and specify a different value as the value to be returned by the program:

$$\text{demoReturn} := \begin{vmatrix} x \leftarrow 3 \\ y \leftarrow 5 \\ \text{return } x \\ z \leftarrow 7 \end{vmatrix}$$

$$\text{demoReturn} = 3$$

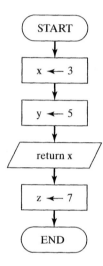

In this program, the `return x` statement says to return the value of x, which is 3. Without the `return` statement, the value on the last line (the value of z, or 7) would be returned.

A `return` statement can be used to set the return value from anywhere in the program, which can be useful when there are numerous loops and/or `if` statements.

Returning Multiple Values

You can return multiple values from a program by returning them as an array. For example, if you wanted to return each of the computed areas from the `CylinderArea` program, you would add an array definition on the (new) last line of the program:

$$\text{CylinderArea}(D, L) := \begin{vmatrix} R \leftarrow \dfrac{D}{2} \\ A_{circle} \leftarrow \pi \cdot R^2 \\ A_{side} \leftarrow \pi \cdot D \cdot L \\ A_{surface} \leftarrow 2 \cdot A_{circle} + A_{side} \\ \begin{pmatrix} A_{circle} \\ A_{side} \\ A_{surface} \end{pmatrix} \end{vmatrix}$$

When the program is used, all three computed areas will be returned as a vector:

$$D := 3 \cdot cm \qquad L := 7 \cdot cm$$

$$\text{CylinderArea}(D, L) = \begin{pmatrix} 7.069 \\ 65.973 \\ 80.111 \end{pmatrix} cm^2$$

Section 6.5 Basic Elements of Programming

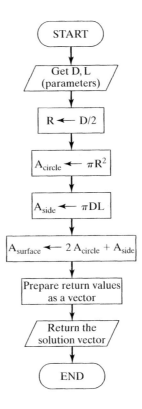

If you want the results available individually after the program is run, assign the returned values to elements of a vector by placing a vector of variables on the left side of an assignment operator (:=).

$$D := 3 \cdot cm \quad L := 7 \cdot cm$$

$$\begin{pmatrix} A_{one} \\ A_{two} \\ A_{three} \end{pmatrix} := \text{CylinderArea}(D, L)$$

$$A_{one} = 7.069 \, cm^2$$

Example: Linear regression of a data set.

What follows is an example of using a Mathcad program to perform a series of operations on a data set and return the results as a vector. This program will perform a linear regression (using Mathcad's built-in functions) on x and y values stored in two vectors, and return the slope, intercept, and R^2 values. First, the data vectors are defined:

$$x := \begin{pmatrix} 1 \\ 2 \\ 3 \\ 4 \\ 5 \end{pmatrix} \quad y := \begin{pmatrix} 2 \\ 5 \\ 8 \\ 13 \\ 17 \end{pmatrix}$$

We could calculate the slope, intercept, and R^2 value using three of Mathcad's built-in functions:

```
sl  := slope(x,y)          sl  = 3.8
int := intercept(x,y)      int = -2.4
R2  := corr(x,y)²          R2  = 0.989
```

Or, alternatively, we can combine these three steps into a program that calculates all three results at the same time:

$$\text{Regress}(xx, yy) := \begin{vmatrix} sl \leftarrow \text{slope}(xx, yy) \\ int \leftarrow \text{intercept}(xx, yy) \\ R2 \leftarrow \text{corr}(xx, yy)^2 \\ \begin{pmatrix} sl \\ int \\ R2 \end{pmatrix} \end{vmatrix}$$

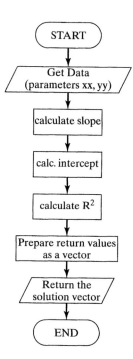

The three computed results are returned as a vector containing three values:

$$\text{soln} := \text{Regress}(x, y) \qquad \text{Soln} = \begin{pmatrix} 3.8 \\ -2.4 \\ 0.989 \end{pmatrix}$$

Note: The vector on the fourth line of the program, containing `sl`, `int`, and `R2`, was created as follows:

1. Add Line was used to create a placeholder in the fourth line of the program.
2. After clicking on the placeholder in the fourth row of the program, a vector of placeholders with three rows and one column was created using the Insert Matrix dialog, [CTRL-M].

3. The three placeholders in the vector were filled with the program's local variable names: `s1`, `int`, and `R2`.

While there is little need to build the `slope()`, `intercept()`, and `corr()` functions into a program to handle a single data set, it might be convenient to have a program like `Regress()` if you need to perform a linear regression on many sets of data.

6.5.5 Conditional Execution

It is extremely important for a program to be able to perform certain calculations under specific conditions. For example, in order to determine the density of water at a specific temperature and pressure, you first have to determine if the water is a solid, a liquid, or a gas at those conditions. A program would use *conditional execution* statements to select the appropriate equation for density.

If Statement

The classic conditional execution statement is the `if` statement. An `if` statement is used to select from two options, depending on the result of a calculated (logical) *condition*. In the following example, the temperature is checked to see if freezing is a concern:

```
checkForIce(Temp) := | RV ← "No Problem"
                     | RV ← "Look Out for Ice!" if Temp < 273.15

checkForIce(280·K) = "No Problem"
checkForIce(250·K) = "Look Out For Ice!"
checkForIce(465·R) = "Look Out For Ice!"
```

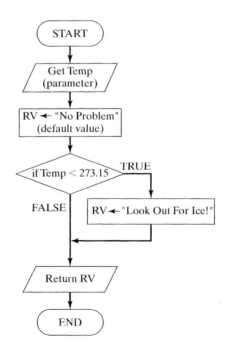

This example also illustrates that you can use units and text strings with Mathcad programs.

To add an `if` statement to a program, click on the placeholder where you want the statement to be placed, and either click on `if` on the Programming toolbar, or press the right-brace key, [}].

Note: You cannot simply type "if" into the placeholder.
When the `if` statement is inserted, there will be a placeholder on both sides of the `if`:

$$\text{checkForIce(Temp)} := \begin{vmatrix} \text{RV} \leftarrow \text{"No Problem"} \\ \blacksquare \text{ if } \blacksquare \end{vmatrix}$$

The placeholder on the right is for the *condition* that will be evaluated; it must evaluate to true or false. The placeholder on the left will contain the program code that should execute if the condition evaluates to true. In this example, we use the left placeholder to assign the string `"Look Out For Ice!"` to the return variable, RV, if the condition in the right placeholder, Temp $<$ 237.15 K, evaluates to true:

$$\text{checkForIce(Temp)} :=$$

$$\begin{vmatrix} \text{RV} \leftarrow \text{"No Problem"} \\ \text{RV} \leftarrow \text{"Look Out For Ice!" if Temp} < 273.15\text{K} \end{vmatrix}$$

APPLICATION: DETERMINING THE CORRECT KINETIC ENERGY CORRECTION FACTOR, α, FOR A PARTICULAR FLOW

The mechanical energy balance is an equation that is commonly used by engineers for determining the pump power required to move a fluid through a piping system. One of the terms in the equation accounts for the change in kinetic energy of the fluid, and includes a *kinetic energy correction factor*, α. The value of α is 2 for fully developed *laminar flow*, and approximately 1.05 for fully developed *turbulent flow*. In order to determine if the flow is laminar or turbulent, we must calculate the *Reynolds number*, defined as

$$\text{Re} = \frac{DV_{avg}\rho}{\mu}$$

where

    D is the inside diameter of the pipe
    V_{avg} is the average fluid velocity
    ρ is the density of the fluid
    μ is the absolute viscosity of the fluid at the system temperature

If the value of the Reynolds number is 2100 or less, then we will have laminar flow. If it is at least 6000, we will have turbulent flow. If the Reynolds number is between 2100 and 6000, the flow is in a transition region and the value of α cannot be determined precisely. We can write a short Mathcad program to first calculate the Reynolds number and then use two `if` statements to set the value of α, depending on the value of the Reynolds number:

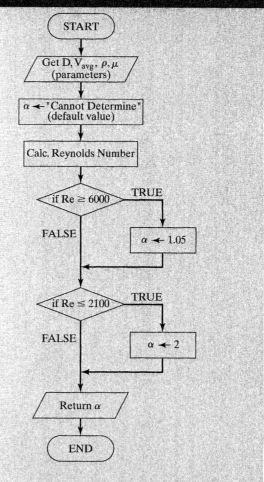

$$\text{setAlpha}(D, V_{avg}, \rho, \mu) := \begin{vmatrix} \alpha \leftarrow \text{"Cannot Determine"} \\ Re \leftarrow \dfrac{D \cdot V_{avg} \cdot \rho}{\mu} \\ \alpha \leftarrow 1.05 \text{ if } Re \geq 6000 \\ \alpha \leftarrow 2 \text{ if } Re \leq 2100 \\ \alpha \end{vmatrix}$$

$$\text{setAlpha}\left(2 \cdot \text{in}, \ 3 \cdot \dfrac{m}{s}, \ 950 \cdot \dfrac{kg}{m^3}, \ 0.012 \cdot \text{poise}\right) = 1.05$$

In the preceding example, the flow of a fluid with a density of 950 kg/m³ and a viscosity of 0.012 poise in a 2-inch pipe at an average velocity of 3 m/s was found to be turbulent, so $\alpha = 1.05$. (The Reynolds number is 120,600.)

Notice that the value of α was initially assigned the text string "Cannot Determine". This is the default case; if both of the `if` statements evaluate to false, the returned α value will be the warning text string. Next, the Reynolds number is determined. Then in program line 3, an `if` statement is used to see if the Reynolds number is less than or equal to 2100. If it is, then the flow is laminar and α is given a value of 2. Line 4 then checks to see if the flow is turbulent and, if so, sets $\alpha = 1.05$. The final line indicates that the value of α is returned by the program since, by default, the value on the last line of the program is returned.

If the velocity is decreased to 0.1 m/s then the Reynolds number falls below 6000 and the program indicates this by sending back the warning string

$$\text{setAlpha}\left(2 \cdot \text{in}, \ 0.1 \cdot \dfrac{m}{s}, \ 950 \cdot \dfrac{kg}{m^3}, \ 0.012 \cdot \text{poise}\right) = \text{"Cannot Determine"}$$

Otherwise Statement

The otherwise statement is used in conjunction with an `if` statement when you want the program to do something when the condition in the `if` statement evaluates to false. For example, we might rewrite the `CheckForIce()` program to test for temperatures below freezing, but provide an `otherwise` to set the text string when freezing is not a concern:

$$\text{checkForIce}(\text{Temp}) := \begin{vmatrix} RV \leftarrow \text{"Look Out For Ice!"} \text{ if Temp} < 273.15K \\ RV \leftarrow \text{"No Problem"} \text{ otherwise} \end{vmatrix}$$

checkForIce(280·K) = "No Problem"
checkForIce(250·K) = "Look Out For Ice!"

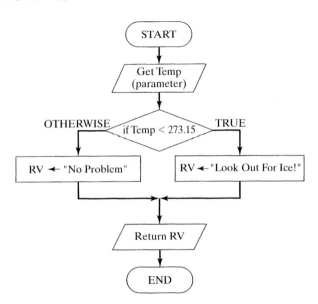

This version is functionally equivalent to the earlier version, but might be more easily read by someone unfamiliar with programming. In the earlier version, the return value was set to "No Problem" and was then overwritten by "Look Out For Ice!" if the temperature was below freezing. In this version of the program, there is no overwriting; the return value is set to one text string or the other, based on the result of the if statement.

Note: The otherwise statement must be inserted using the Programming toolbar, or by pressing [Ctrl- }]. You cannot simply type "otherwise" into the placeholder.

On Error Statement

The on error statement is used for *error trapping*, which provides an alternative calculation path when certain values are known to cause errors in a program. When you include an on error statement on a program line, there is a placeholder on both sides of the statement. The placeholder on the right is for the normal calculation, assuming that there is no error. The placeholder on the left is for the alternative calculation in case an error occurs. For example, an inverseValue(x) function will fail if the value of x is zero. However, we can trap this error and return ∞ (evaluates to 1×10^{307}) instead:

$$\text{inverse Value(x)} := \begin{vmatrix} \text{"error trapping example"} \\ \infty \text{ on error } \frac{1}{x} \end{vmatrix}$$

inverseValue(4) = 0.25
inverseValue(0) = 1×10^{307}

Note: The on error statement must be inserted using the Programming toolbar, or by pressing [Ctrl-']. You cannot simply type "on error" into the placeholder.

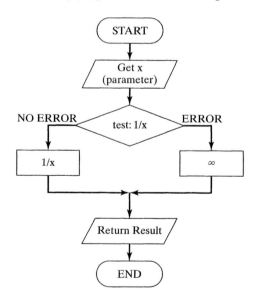

6.5.6 Loops

Loop structures are used to cause programs to perform calculations over and over again. There are several instances when these repetitive calculations are desirable:

- if you want to repeat a series of calculations for each value in a data set or matrix.
- if you want to perform an iterative (guess-and-check) calculation until the guessed and calculated values are within a preset tolerance.

- if you want to move through the rows of data in an array until you find a value that meets a particular criterion.

There are two loop structures supported in Mathcad programs: *while loops* and *for loops*.

While Loops

A `while` loop is a control structure that causes an action to be repeated (iteration) while a condition is true. As soon as the condition is false, the iteration stops. In the following example, the loop continues to operate as long as the value of local variable x is less than 100:

$$\text{demo While(seed)} := \begin{vmatrix} x \leftarrow \text{seed} \\ \text{while } x < 100 \\ \quad x \leftarrow x^{2.3} \\ x \end{vmatrix}$$

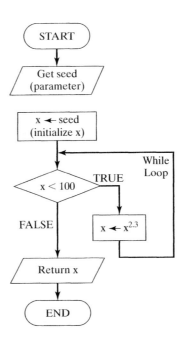

When x is greater than or equal to 100, the loop terminates and the final value of x is returned.

Note: You must use the Programming toolbar or press [CTRL+]] to create the `while` loop. You cannot simply type "while" in a placeholder.

The `seed` variable can be assigned any value, but some may cause problems, as illustrated in the following code:

```
demoWhile(2)   = 4.599 × 10³
demoWhile(10)  = 199.526
demoWhile(200) = 200
demoWhile(-3)  = ∎
```

In this example, seed values between 0 and 1 will put the `while` loop into an infinite cycle, and negative values are invalid in the $x^{2.3}$ statement. If a loop is running indefinitely you can press the escape key [ESC] to terminate your program. When a program statement contains an error that prevents it from running, Mathcad indicates that the program failed by showing the equation in red.

For Loops

A for loop uses an *iteration variable* to loop a prescribed number of times. In the following example, the iteration variable is called j and takes on values from 1 to 5 (starting with a value of 1, and increasing by 1 each time through the for loop):

$$\text{demoFor}(k) := \begin{vmatrix} \text{outVar} \leftarrow k \\ \text{for } j \in 1..5 \\ \text{outVar} \leftarrow \text{outVar} + j \\ \text{outVar} \end{vmatrix}$$

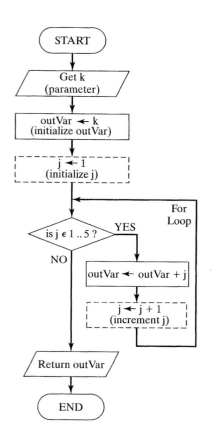

Notes:

1. You must use the Programming toolbar or press [CTRL + "] to create the for loop. You cannot simply type "for" in a placeholder.
2. The ∈ symbol is read "in the range of" and is a reminder that the iteration variable is a range variable, and thus is defined as a range variable. The range 1 .. 5 was entered as [1] [;] [5], where the semicolon was used to enter the ellipsis, "..".
3. In the flowchart for this program, the steps in which j is initialized and incremented are shown with dotted-line borders. This is meant as a reminder

that initialization and incrementation of the index variable, j, is automatically handled by the for loop.

4. The iteration variable can be incremented by values other than 1 by entering the first two values of the range, then entering the ellipsis and the final value. For example, to increment j from 1 to 10 by 2 each time through the loop, use

$$j \in 1, 3 .. 10$$

You can also increment down. For example, to increment j from 10 to 1 using a step size of 1, use

$$j \in 10, 9 .. 1$$

When the demoFor program is run with a k value of 0, it should return a value of $0 + 1 + 2 + 3 + 4 + 5 = 15$:

$$\text{demoFor(k)} := \begin{vmatrix} \text{outVar} \leftarrow \text{k} \\ \text{for } j \in 1..5 \\ \text{outVar} \leftarrow \text{outVar} + j \\ \text{outVar} \end{vmatrix}$$

demoFor(0) = 15

If k = 5, then the return value is 20:

demoFor(5) = 20

Break Statement

The break statement is used to halt execution of a for or while loop. It is used when you want some condition other than the normal loop termination to stop the loop. It can be used to stop a loop that might generate an error, or just to provide another way out of

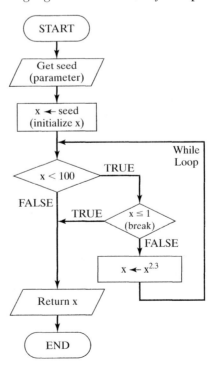

the loop when necessary. For example, we could use a `break` statement to terminate a `while` loop that would otherwise go on indefinitely:

$$\text{demoWhile(seed)} := \begin{vmatrix} x \leftarrow \text{seed} \\ \text{while } x < 100 \\ \quad \begin{vmatrix} \text{break if } x \leq 1 \\ x \leftarrow x^{2.3} \end{vmatrix} \\ x \end{vmatrix}$$

$$\text{demoWhile}(0.7) = 0.7$$

Note: You must use the Programming toolbar or press [CTRL + {] to insert the `break` statement. You cannot simply type "break" in a placeholder.

Continue Statement

A `continue` statement is used with nested loops (loops within other loops) to halt execution of the loop containing the `continue` statement, and continuing the program with the next iteration of the next outer loop. For example, the following program, without a `continue` statement, has three nested `for` loops using iteration variables i, j, and k:

$$\text{demoNoContinue} := \begin{vmatrix} \text{sum} \leftarrow 0 \\ \text{for } i \in 1..6 \\ \quad \text{for } j \in 1..5 \\ \quad\quad \text{for } k \in 1..4 \\ \quad\quad\quad \text{sum} \leftarrow \text{sum} + 1 \\ \text{sum} \end{vmatrix}$$

$$\text{demoNoContinue} = 120$$

The computed `sum` is 120, because the i loop cycled 6 times, the j loop cycled 6 × 5 times, and the k loop cycled 6 × 5 × 4 = 120 times, and the `sum` calculation was inside the k loop.

Now we add a `continue` statement that halts the j loop if j > 3:

$$\text{demoWithContinue} := \begin{vmatrix} \text{sum} \leftarrow 0 \\ \text{for } i \in 1..6 \\ \quad \text{for } j \in 1..5 \\ \quad\quad \text{continue if } j > 3 \\ \quad\quad \text{for } k \in 1..4 \\ \quad\quad\quad \text{sum} \leftarrow \text{sum} + 1 \\ \text{sum} \end{vmatrix}$$

$$\text{demoWithContinue} = 72$$

This time, `sum` has a value of 72, because the i loop cycled 6 times, the j loop cycled 6 × 3 times, and the k loop cycled 6 × 3 × 4 = 72 times.

Notes:

1. You must use the Programming toolbar or press [CTRL + [ ] to insert the `continue` statement. You cannot simply type "continue" in a placeholder.
2. The `continue` statement is typically used with an `if` statement. You must enter the `if` statement first to create the placeholder for the `continue` statement.

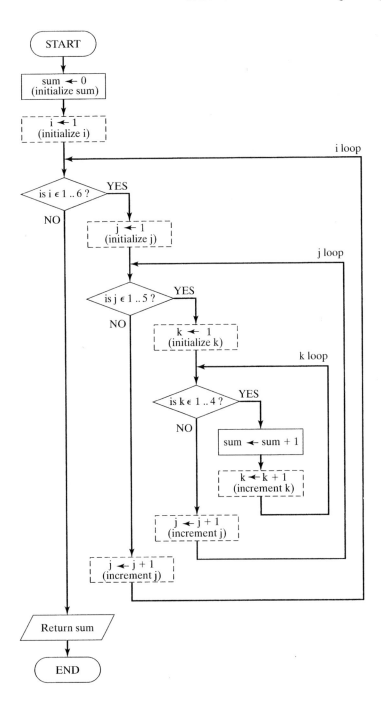

3. In the flowchart for this program, the initialization and incrementation steps that are automatically handled by the for loop have been omitted.

6.5.7 Functions

Functions are an indispensable part of modern programming because they allow a program to be broken down into pieces, each of which ideally handles a single task (i.e., performs a single function). The programmer can then call upon the functions as needed to complete a more complex calculation.

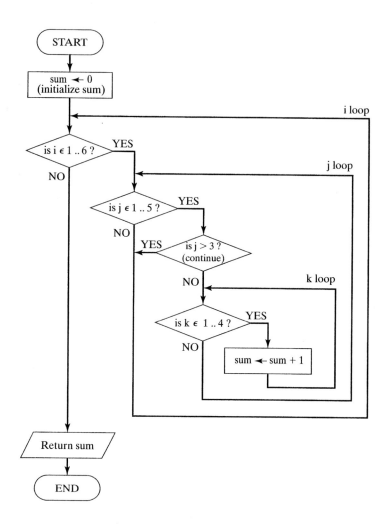

Functions are an indispensable part of Mathcad as well. Mathcad provides built-in functions that can be used as needed to complete a lot of computational tasks. If Mathcad's built-in functions cannot perform a calculation, you can write your own function and use it with Mathcad's built-in functions. A Mathcad program is essentially a multi-step, user-written function.

SUMMARY

Creating a Mathcad Program

Program Name   Identifies the program and is used as a function name when the program is used on the worksheet.

Parameter List   Used to pass values into a program.

Assignment Operator, :=   Connects the program name and parameter list to the program lines, which are indicated by heavy vertical lines.

Add Line   Used to insert lines into the program; the keyboard shortcut is [ ]] (right-square-bracket key.)

Basic Elements of Programming

Data	Data values are stored in Mathcad variables—either as single values or as arrays.
Input	Mathcad programs can receive input from • the worksheet, • the parameter list, or • data files (text files) using the `READPRN()` function.
Operations	Nearly all of Mathcad's standard operators may be used in programs, including the standard math operators and matrix operators listed in the tables that follow. Variables defined within a Mathcad program must be defined with the local definition operator, ←. These variables are only defined within the program region (local scope).
Output	Results from Mathcad programs must be returned from the program. By default, the value on the last assignment statement in the program is the return value. You can override the default by using the `return` statement (from the Programming toolbar) in your program. Multiple values may be returned from Mathcad programs as an array.
Conditional Execution	Mathcad's `if` statement is used for conditional execution—to allow the flow of the program to follow different paths depending on the value of the condition following the `if` statement. The if statement is found on the Programming toolbar or it may be entered using [}] (right brace).
Loops	Mathcad programs support two types of loops: • `for` loops—used when you want the loop to execute a specified number of times. • `while` loops—used when you want the looping to continue as long as a specified condition is true. You can use the `break` statement to terminate a loop or a `continue` statement to cause program flow to jump out of an inside loop, but continue in another loop. Both `break` [Ctrl-{] and `continue` [Ctrl-[ ] are available on the Programming toolbar.
Functions	Mathcad programs are multistep functions and may include other functions, including Mathcad's built-in functions.

Standard Math Operators

SYMBOL	NAME	SHORTCUT KEY
+	Addition	+
−	Subtraction	−
*	Multiplication	[Shift-8]
/	Division	/
e^x	Exponential	
$1/x$	Inverse	
x^y	Raise to a Power	[∧], or [Shift-6]
$n!$	Factorial	!
$|x|$	Absolute Value	
$\sqrt{\ }$	Square Root	\
$\sqrt[n]{\ }$	Nth Root	[Ctrl-\]

Matrix Operators

SYMBOL	NAME	SHORTCUT KEY
+	Addition	+
−	Subtraction	−
*	Multiplication	[Shift-8]
/	Division	/
A_n	Array Element	[
M^{-1}	Matrix Inverse	
$|M|$	Determinant	|
→	Vectorize	[Ctrl- −]
$M^{<x>}$	Matrix Column	[Ctrl-6]
M^T	Matrix Transpose	[Ctrl-1]
$M_1 \cdot M_2$	Dot Product	*, or [Shift-8]
$M_1 \times M_2$	Cross Product	[Ctrl-8]

Operator Precedence Rules

PRECEDENCE	OPERATOR	NAME
First	∧	Exponentiation
Second	*, /	Multiplication, Division
Third	+, −	Addition, Subtraction

It is a good idea to include the parentheses to make the order of evaluation clear.

Flowchart Symbols

SYMBOL	NAME	USAGE
⬭	Terminator	Indicates the start or end of a program
▭	Operation	Indicates a computation step
▱	Data	Indicates an input or output step
◇	Decision	Indicates a decision point in a program.
○	Connector	Indicates that the flowchart continues in another location.

KEY TERMS

argument list
break statement
conditional execution
continue statement
data
flowchart
for loop
function
if statement

input
input source
local definition
loops
on error statement
operations
operator
operator precedence
otherwise statement

output
parameter list
passing by value
program
program region
Programming Toolbar
return statement
return value
while loop

Problems

1. **CHECKING FOR FEVER**

   For a Practice! Exercise in this chapter you were asked to create a flowchart illustrating the decision tree for determining whether a person's fever requires medication, and whether or not the medication should be aspirin. Now, write a Mathcad program that receives the patient's oral temperature and age through the parameter list and returns two text strings, one for each of the following decisions:

   Decision 1: Should the patient receive medication?
   - "patient should receive medicine" or "patient does not require medicine"

   Decision 2: Can the medicine be aspirin?
   - "no medicine required" or "aspirin is OK" or "do not give aspirin"

2. **CALCULATING GRADES**

   a. Write a flowchart for a program that receives a numerical score (0–100) through the parameter list and then determines a letter grade based on the following information:

Score $\geq$ 90	A
80 $\geq$ Score < 90	B
70 $\geq$ Score < 80	C
60 $\geq$ Score < 70	D
Score < 60	F

b. Write a Mathcad program that will receive the numerical score through the parameter list and return the appropriate letter grade.

3. CALCULATING AVERAGE AND MEDIAN SCORES

   a. Write a program that receives a vector of scores through the parameter list, then uses Mathcad's mean() and median() functions to compute the average and median scores.

   b. Test your program with the following values:

   $$\text{Scores} := \begin{pmatrix} 98 \\ 95 \\ 92 \\ 94 \\ 90 \end{pmatrix} \qquad \begin{array}{l} \text{mean}(\text{Scores}) = 93.8 \\ \text{median}(\text{Scores}) = 94 \end{array}$$

   c. Use your program to determine the average and median scores for the following data set:

   $$\text{ExamScores} := \begin{pmatrix} 58 \\ 92 \\ 45 \\ 84 \\ 93 \\ 60 \\ 91 \\ 55 \\ 97 \end{pmatrix}$$

4. RESOLUTION OF A FORCE VECTOR

   A common calculation in physics and engineering mechanics is the determination of the horizontal and vertical components of a force acting at an arbitrary angle. For example, a 250-N force acting at 150° (from 3 o'clock, which is called zero degrees) has a horizontal component of −216.5 N (F cos(150°)) and a vertical component of 125 N (F sin(150°)):

   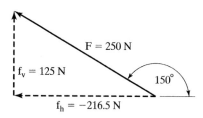

   a. Write a Mathcad program that receives the magnitude and direction (angle) of a force, and returns both the horizontal and vertical components.
   b. Test your program using the 250-N force example.
   c. Use your program to determine the horizontal and vertical components of the following forces:
      i.   250 N at 60°.
      ii. 1200 N at 220°
      iii. 840 lb$_f$ at 5 radians

5. **QUALITY CONTROL IN A BOTTLING PLANT, I**

   The manager of a bottling plant wants to keep profits as high as possible while keeping customer complaints low. The volume of each bottle is measured automatically, and after every 10 bottles the volumes are recorded.

   a. Write a Mathcad program that will receive a vector of 10 volume values and then use Mathcad's mean() and stdev() functions to calculate the average and standard deviation of the volume values. Have your program return both calculated values.

   b. Test your program with the following values:

   $$\text{bottles} := \begin{pmatrix} 201.5 \\ 202.3 \\ 203.4 \\ 202.1 \\ 200.5 \\ 203.1 \\ 201.1 \\ 202.0 \\ 201.4 \\ 201.1 \end{pmatrix} \qquad \begin{array}{l} \text{mean(bottles)} = 201.85 \\ \text{stdev(bottles)} = 0.867 \end{array}$$

   c. Use your program to determine the average and standard deviation of the following volume values:

   $$\text{largeBottles} := \begin{pmatrix} 503 \\ 497 \\ 512 \\ 502 \\ 517 \\ 505 \\ 499 \\ 501 \\ 511 \\ 482 \end{pmatrix}$$

6. **QUALITY CONTROL IN A BOTTLING PLANT, II**

   The manager of another bottling plant wants to keep profits as high as possible while keeping customer complaints low. The volume of each bottle is measured automatically, and after every 10 bottles the volumes are recorded. She wants each bottle to contain at least 200 ml, but not more than 204 ml, the average volume to be less that 202 ml, and the standard deviation of the 10 measurements to be less than 1.5 ml.

   a. Write a flowchart of a program that will
      - receive a vector of 10 volume values,
      - use a for loop to check the individual volumes of each bottle,
      - use Mathcad functions to determine the average volume and standard deviation for the 10 values, and
      - return a text string that indicates whether the bottles were or were not filled correctly. If they were not, have the program indicate (with a text string) which of the criteria is/are not being met.

b. Write a Mathcad program that implements your flowchart. Test the program with the following values:

$$\text{bottles} := \begin{pmatrix} 201.5 \\ 202.3 \\ 203.4 \\ 202.1 \\ 200.5 \\ 203.1 \\ 201.1 \\ 202.0 \\ 201.4 \\ 201.1 \end{pmatrix} \qquad \begin{aligned} \text{mean(bottles)} &= 201.85 \\ \text{stdev(bottles)} &= 0.867 \end{aligned}$$

c. Create a new set of test data that will allow you to demonstrate that your `for` loop is working correctly to check the volume of each bottle.

d. Use your program to see if the bottles represented by the following values were filled correctly:

$$\text{moreBottles} := \begin{pmatrix} 201.3 \\ 201.7 \\ 203.1 \\ 201.8 \\ 203.5 \\ 203.3 \\ 202.1 \\ 201.0 \\ 201.6 \\ 202.1 \end{pmatrix}$$

e. How would your program have to be modified if the sample size was increased from 10 bottles to 100?

7. **USING A FOR LOOP TO COUNT OCCURRENCES, I**

Write a program that uses a `for` loop to count the number of exam scores in the 80s.

$$\text{Scores} := \begin{pmatrix} 85 \\ 82 \\ 93 \\ 90 \\ 57 \\ 84 \\ 91 \\ 88 \\ 79 \\ 82 \end{pmatrix}$$

Note: Mathcad provides the `last(v)` function to return the index value of the last element in vector, v. This is an easy way to set the limit of the `for` loop.

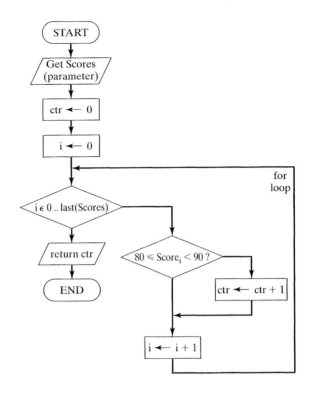

8. **USING A FOR LOOP TO COUNT OCCURRENCES, II**

   One approach to quality control is to improve product quality by reducing product variability. For example, an extruded plastic part might have a width that varies somewhat. The parts will fit better with the rest of the system if the variability in width is tightly controlled.

   a. Write a program that uses a `for` loop to count the number of parts that are outside of the width specification range between 0.995 cm and 1.004 cm:

   $$\text{width} := \begin{pmatrix} 1.011 \\ 0.997 \\ 0.999 \\ 0.993 \\ 1.004 \\ 0.998 \\ 1.005 \\ 0.999 \\ 1.000 \\ 0.998 \\ 0.989 \\ 0.990 \\ 0.986 \\ 0.994 \\ 0.998 \\ 1.001 \end{pmatrix}$$

   b. Create a flowchart showing how your program functions.

9. **USING A WHILE LOOP TO DETERMINE TIME BELOW A CRITICAL DEPTH**

   Scuba divers need to be careful about how much time they spend at a certain depth because of the potentially harmful buildup of soluble gases in their blood (a condition called "the bends"). Wrist-mounted depth meters can continuously record a diver's depth throughout the dive, and the data can be used to determine the time the diver spent below a critical depth, such as 35 feet. The following table is illustrative:

DEPTH (ft)	TIME (min.)	DEPTH (ft)	TIME (min.)	DEPTH (ft)	TIME (min.)
0	0	43	10	34	20
5	1	42	11	30	21
8	2	44	12	28	22
12	3	41	13	25	23
15	4	41	14	21	24
21	5	43	15	17	25
30	6	45	16	9	26
38	7	44	17	7	27
42	8	42	18	4	28
41	9	37	19	0	29

   a. Write a Mathcad program that uses a `while` loop to determine the length of time that a diver spent below 35 feet.

   b. Create a flowchart showing how your program functions.

10. **DIRECT SUBSTITUTION ITERATIVE METHOD**

    One simple iterative solution technique is a method called *direct substitution*, shown in the flowchart. The method requires that the function you are trying to solve be written with an unknown on both sides of the equation, and the unknown alone on the left side of the equation. For example, consider the following equation with two obvious roots $(x = 4, x = 7)$:

    $$(x - 4)(x - 7) = 0$$

    This can be written as a polynomial:

    $$x^2 - 11x + 28 = 0$$

    We can rearrange this equation so that there is an x by itself on the left side:

    $$x = \frac{x^2 + 28}{11}$$

    To solve this equation by the *direct substitution* method, you use your guess value (call it x_G) on the right side of the equation, and solve for a calculated

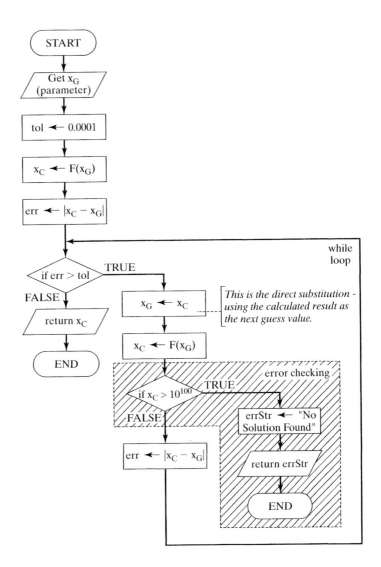

value (x_C) on the left:

$$x_C = \frac{x_G^2 + 28}{11}$$

The F (x_G) that appears in the flowchart is the right side of this equation and is used to compute x_C:

$$F(X_G) = \frac{x_G^2 + 28}{11}$$

This function must be defined on your Mathcad worksheet before the direct substitution program is used.

a. Write a dirSub(x_G) program based on the flowchart. For now, skip the error-checking section to keep the program a little simpler.
b. Use the test function F(x_G) shown previously to test your program. Try various initial-guess values, x_G. Can you obtain either of the two solutions ($x = 4, x = 7$)?

c. The direct substitution method cannot find all solutions, and it should fail to find one of the solutions to the test function (unless you enter the solution as the guess value). When the program fails to find a solution, it is said to diverge from, rather than converge on, the solution. How does Mathcad indicate that there is a divergence problem?

d. Now add the error-checking section of the flowchart to your program. With the error-checking code in place, how does the program respond if the method diverges?

e. Try a polynomial without an obvious solution, $0.3x^2 - 9x + 24 = 0$. Rearrange the equation as required for the direct substitution method, and use your program to search for roots. (You can use the `polyroots()` function to check your answer.)

7

Mathcad's Symbolic Math Capabilities

COMPOSITE MATERIALS

The field of *composite materials* has already had an impact on most of us in the developed world. The sporting goods industry, for example, has found that lighter and stronger composite materials allow athletes to go faster and farther. Composites are common in modern aircraft for the same reasons. Still, the field of composite materials really is just getting started.

Material scientists have long analyzed the behavior of materials under various stresses. Concrete, for example, is known to perform well under compression, but doesn't hold up under tension. Cables are designed for holding loads under tension, but not under compression. (You can't push a rope!) The analytical skills of the materials scientists are required to understand composite materials, but they now go a step further: It has become possible to talk about designing materials for specific purposes. If a part needs to perform better under tension, you might change the type of fibers used in the composite. If the part needs better resistance to compression, the type of matrix used to surround the fibers might be changed. For better bending performance, the bonding between the fibers and the matrix might be improved. But in order to design a better composite material, you have to know how each element of the

SECTIONS

7.1 Symbolic Math Using Mathcad
7.2 Solving an Equation Symbolically
7.3 Manipulating Equations
7.4 Polynomial Coefficients
7.5 Symbolic Matrix Math
7.6 Symbolic Integration
7.7 Symbolic Differentiation

OBJECTIVES

After reading this chapter, you will

- be able to solve for a variable in an equation using Mathcad's symbolic math processor
- be aware that Mathcad has the ability to perform standard symbolic math operations, such as substitutions, factoring, and expanding terms
- know that Mathcad can perform matrix operations symbolically
- be able to use Mathcad's symbolic math capabilities to integrate and differentiate functions

composite works, as well as how all the elements interact with each other. You also need to understand the chemical and physical natures of the elements of the composite, in addition to the properties imparted to the elements and the final composite by the manufacturing processes. This is a field that will require the skills of a wide range of scientists and engineers working together.

7.1 SYMBOLIC MATH WITH MATHCAD

Scientists and engineers commonly want a numerical value as the result of a calculation, and Mathcad's numerical math features provide this type of result. However, there are times when you need a result in terms of the mathematical symbols themselves, and you can use Mathcad's *symbolic math* capability for that type of calculation. A symbolic result may be required when

- you want to know how Mathcad obtained a numerical result by seeing the equation solved symbolically,
- you want greater precision on a matrix inversion, so you have Mathcad invert the matrix symbolically, rather than numerically, or
- you want to integrate a function symbolically and use the result in another function.

Mathcad provides symbolic math functions in two locations: the *Symbolics menu* at the top of the window and the *Symbolic Keyword Toolbar*. Both locations provide access to essentially the same symbolic capabilities, with one important distinction: The Symbolic Keyword Toolbar uses the *live symbolic operator*, or $\rightarrow$ symbol. This is called a "live" operator because it automatically recalculates the value of an expression whenever information to the left of or above the operator changes.

The results of calculations carried out using the Symbolics menu are not live. That is, once a symbolic operation has been performed using commands from the Symbolics menu, the result will not be automatically updated, even if the input data change. Thus, to update a result using the Symbolics menu, you must repeat the calculation.

The Symbolics menu is a bit more straightforward than the Symbolic Keyword Toolbar for solving for a particular variable and for factoring an expression. The Symbolic Keyword Toolbar is simpler for substitution. Both approaches will be demonstrated.

Symbolics Menu

The Symbolics menu is on the menu bar at the top of the Mathcad window. One of the first considerations in using the symbolic commands is how you want the results to be displayed. You can set this feature by using the Evaluation Style dialog box from the Symbolics menu (shown on the next page). The Evaluation Style dialog box controls whether the results of a symbolic operation are presented to the right of the original expression or below the original expression. When results are placed below the original expression, the new equation regions can start running into existing regions. This can be avoided by having Mathcad insert blank lines before displaying the results. By default, Mathcad places the results below the original expression and adds blank lines to avoid overwriting other equation regions.

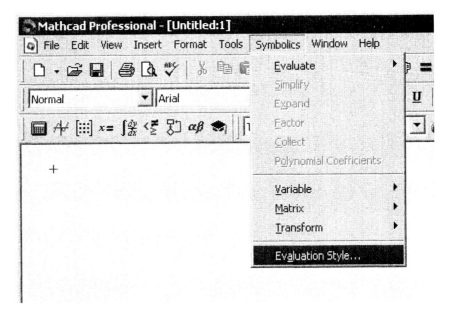

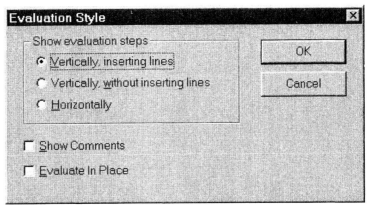

Note: If you select "Evaluate in Place," your original expression will be replaced by the computed result.

As an example of using the Symbolics menu, we will find the solutions of the equation:

$$(x - 3) \cdot (x - 4) = 0.$$

This equation has two obvious solutions (x = 3, x = 4), so it is easy to see whether Mathcad is finding the correct ones.

PROFESSIONAL SUCCESS

Keep test expressions as simple as possible.
When you are choosing mathematical expressions to test the features of a software package or to validate your own functions, try to come up with a test that

- is just complex enough to demonstrate that the function is (or is not) working correctly, and
- has an obvious, or at least a known, solution, so that it is readily apparent whether the test has succeeded or failed.

Since we want to find the values of x that satisfy the equation, click on the equation and select either one of the x variables. Note that symbolic equality (=) was used in the equation:

$$(x-3) \cdot (x-4) = 0$$

To solve for the variable x, use Symbolics/Variable/Solve:

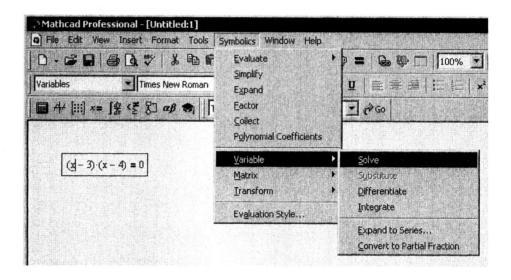

The solutions—the two values of x that satisfy the equation—are presented as a two-element vector:

$$(x-3) \cdot (x-4) = 0$$

$$\begin{bmatrix} 3 \\ 4 \end{bmatrix}$$

Symbolic operations are often performed on expressions, rather than complete equations. If you solve the expression $(x - 3) \cdot (x - 4)$ for x, Mathcad will set the expression to zero before finding the solutions. Thus, we would have

$$(x-3) \cdot (x-4)$$

$$\begin{bmatrix} 3 \\ 4 \end{bmatrix}$$

Symbolic Keyword Toolbar

The Symbolic Keyword Toolbar is available from the Math Toolbar. Click on the button showing a mortarboard icon to open the Symbolic Keyword Toolbar. When opened, the Symbolic Keyword Toolbar looks like this:

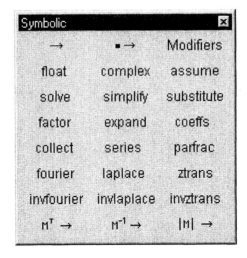

To use the features on this Toolbar, first enter an expression, and then select the expression and click one of the buttons on the Toolbar to perform a symbolic operation on the expression. For instance, the preceding example, we know that the equation

$$(x - 3) \cdot (x - 4) = 0$$

has two solutions: $x = 3$ and $x = 4$. Mathcad can also find those solutions using the [solve] button on the Symbolic Keyword Toolbar. This will be demonstrated in the next section.

7.2 SOLVING AN EQUATION SYMBOLICALLY

If you enter an incomplete equation, such as the left-hand side of the previous equation, Mathcad will set the expression equal to zero when the solve operation is performed. To solve the equation in this manner, enter the expression to be solved and select it:

$$(x - 3) \cdot (x - 4)$$

Then click on the [solve] button on the Symbolic Keyword Toolbar:

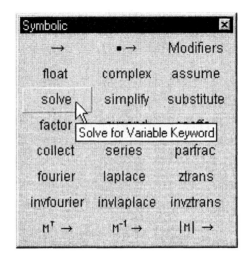

The word "solve" appears after the expression, with an empty placeholder. Mathcad needs to be told which variable to solve for. Enter an x in the placeholder and then press [enter]:

```
(x-3)·(x-4)  solve, ▮ →
```

Mathcad returns the result. In this case, since there are two solutions, the solutions are returned as a two-element vector:

$$(x-3)\cdot(x-4) \quad \text{solve, } x \rightarrow \begin{bmatrix} 3 \\ 4 \end{bmatrix}$$

Note: When using the Symbolic Keyword Toolbar, you tell Mathcad which variable to solve for by typing the variable name into the placeholder. When using the Symbolics menu to solve for x, you tell Mathcad to solve for x by first selecting one of the x variables in the expression and then choosing "solve" from the menu.

If Mathcad cannot solve an expression symbolically, nothing will be displayed on the right side of the arrow operator, and the expression will be displayed in red to indicate that an error has occurred:

```
(x-3)·(x-4)  solve,y →
```

If you click on the expression, Mathcad will display an error message indicating that no solution was found.

PRACTICE!

Use Mathcad to solve the following equations (you can leave off the zero when entering the expressions into Mathcad):

a. $(x-1)\cdot(x+1) = 0$.
b. $x^2 - 1 = 0$.
c. $(y-2)\cdot(y+3) = 0$.
d. $3z^2 - 2z + 8 = 0$.
e. $\dfrac{x+4}{x^2 + 6x - 2} = 0$.

7.3 MANIPULATING EQUATIONS

A number of algebraic manipulations, such as factoring out a common variable, are frequently used when one is working with algebraic expressions, and Mathcad implements these through the Symbolic Keyword Toolbar. These routine manipulations include the following:

- *Expanding* a collection of variables (e.g., after factoring).
- *Factoring* a common variable out of a complex expression.
- *Substituting* one variable or expression for another variable.
- *Simplifying* a complex expression.
- *Collecting terms* on a designated variable.

A *partial-fraction expansion* in which a complex expression is expanded into an equivalent expression consisting of products of fractions is a somewhat more complex operation, but one that can be helpful in finding solutions. Although this procedure will not work on all fractional expressions, it can be very useful in certain circumstances.

Examples of each of the preceding manipulations follow.

Expand

Expanding the expression $(x-3) \cdot (x-4)$ yields a polynomial in x. To expand an expression using the Symbolics menu, select the expression, and then choose Symbolics/Expand:

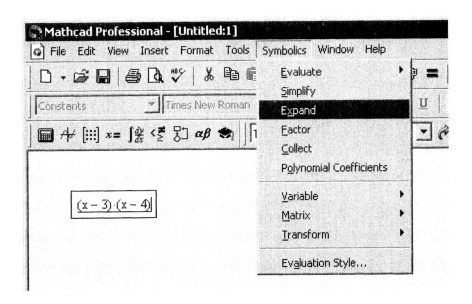

After expansion, the result is placed below the original expression

$(x-3) \cdot (x-4)$
$x^2-7 \cdot x+12$

Alternatively, select the expression, and then press the [expand] button on the Symbolic Keyword Toolbar. In the placeholder, tell Mathcad to expand on x:

$(x-3) \cdot (x-4)$ expand, $x \rightarrow x^2-7 \cdot x+12$

Factor

Factoring reverses the expand operation, pulling an x out of the polynomial:

$x^2-7 \cdot x+12$ factor, $x \rightarrow (x-3) \cdot (x-4)$

Probably a more common usage of the factor operation is to pull a similar quantity out of a multiterm expression. For example, $2\pi r$ appears in both terms in the expression for the surface area of a cylinder. To factor this expression, select the terms to be factored (the entire expression in this case), and then use Symbolics/Factor. The result is placed below the original expression:

$2 \cdot \pi \cdot r^2 + 2 \cdot \pi \cdot r \cdot L$
$2 \cdot r \cdot \pi \cdot (r+L)$

You can also factor only a portion of an expression. For example, we could factor only the right-hand side of the following equation:

$$\boxed{A_{cyl} = \underline{2 \cdot \pi \cdot r^2 + 2 \cdot \pi \cdot r \cdot L}}$$

To factor only the right-hand side, select that side, and then use Symbolics/Factor:

$$A_{cyl} = 2 \cdot r \cdot \pi \cdot (r+L)$$

PRACTICE!

> Try using the Symbolics menu to factor the x out of these expressions:
> a. $6x^2 + 4x$.
> b. $3xy + 4x - 2y$. (Select only part of this expression before factoring.)

Substitute

The substitute operation replaces a variable by another expression. Because substitution is a bit simpler using the Symbolic Keyword Toolbar buttons, that approach will be shown first.

7.3.1 Substitution Using the Symbolic Keyword Toolbar

A simple replacement, such as replacing all the x's in an expression with y's, is easily carried out using the [substitute] button on the Symbolic Keyword Toolbar:

$$(x-3) \cdot (x-4) \quad \text{substitute,} \ x = y \rightarrow (y-3) \cdot (y-4)$$

But you can also replace a variable with a more complicated expression. For example, suppose you wanted to replace the x's with an exponential expression, such as $e^{-t/\tau}$. The [substitute] button can handle this; simply enter the complete exponential expression into the placeholder in the substitute command:

$$(x-3) \cdot (x-4) \quad \text{substitute,}$$
$$x = e^{\frac{-t}{\tau}} \rightarrow \left(\exp\left(\frac{-t}{\tau}\right) - 3\right) \cdot \left(\exp\left(\frac{-t}{\tau}\right) - 4\right)$$

7.3.2 Substitution Using the Symbolics Menu

If you want to use the Symbolics menu to carry out a substitution, there are two things to keep in mind:

1. The new expression (the expression that will be substituted into the existing expression) must be copied to the Windows clipboard before performing the substitution.
2. You must select the variable to be replaced before performing the substitution.

To repeat the last example, the $e^{-t/\tau}$ would be entered into the Mathcad worksheet, selected, and copied to the Windows clipboard, using Edit/Copy:

$$\boxed{e^{\frac{-t}{\tau}}}$$

Then one of the x variables in $(x-3) \cdot (x-4)$ would be selected. (This tells Mathcad to replace all of the x's in the expression with the contents of the Windows clipboard.)

Finally, the substitution is performed using Symbolics/Variable/Substitute, and the results are placed below the original expression. The final Mathcad worksheet now looks like this (with the added comments):

$$e^{\frac{-t}{\tau}}$$ *entered on worksheet, then copied to Windows clipboard*

$$(x-3) \cdot (x-4)$$ *one x selected before substitution*

$$\left(\exp\left(\frac{-t}{\tau}\right) - 3\right) \cdot \left(\exp\left(\frac{-t}{\tau}\right) - 4\right)$$ *the result of the substitution*

Simplify

According to the Mathcad help files, the Simplify menu command "performs arithmetic, cancels common factors, uses basic trigonometric and inverse function identities, and simplifies square roots and powers." If we try to simplify $(x-3) \cdot (x-4)$ using Symbolics/Simplify, the expression is returned unchanged: Mathcad thinks that $(x-3) \cdot (x-4)$ is as simple as this expression gets. In order to demonstrate the Simplify operation, we need to complicate the example a little. Consider this modification:

$$(x-\sqrt{9}) \cdot (x-2^2)$$

If we try to simplify this expression using Symbolics/Simplify, the original expression is returned:

$$(x-\sqrt{9}) \cdot (x-2^2)$$

$$(x-3) \cdot (x-4)$$

The Simplify operation simplified the square root (selecting the positive root) and power. Perhaps a more significant use of the operation is obtaining a common denominator. For example, by selecting the entire expression

$$\frac{a}{(x-3)} + \frac{b}{(x-4)}$$

and then using the Symbolics/Simplify function, the terms will combine over a common denominator:

$$\frac{(a \cdot x - 4 \cdot a + b \cdot x - 3 \cdot b)}{((x-3) \cdot (x-4))}$$

You can also factor the numerator to see the process used to obtain the common denominator. First, you factor the a out of the first two terms in the numerator by selecting those two terms:

$$\frac{(a \cdot x - 4 \cdot a) | + b \cdot x - 3 \cdot b)}{[(x-3) \cdot (x-4)]}$$

Then, you use Symbolics/Factor. Next, you choose the last two terms in the new numerator:

$$\frac{(a \cdot x - 4 \cdot a + b \cdot x - 3 \cdot b|)}{[(x-3) \cdot (x-4)]}$$

Again, you use Symbolics/Factor. The final result is

$$\frac{(a \cdot (x-4) + b \cdot (x-3))}{((x-3) \cdot (x-4))}$$

PRACTICE!

Use the Symbolics menu to simplify these expressions:

a. $\dfrac{x}{x+4} - \dfrac{12}{x+2}$

b. $\dfrac{1}{x} + \dfrac{x}{x-7} - \dfrac{x+6}{x^2}$

c. $\sqrt{\dfrac{4x}{y^2}}$

Collect

The Collect operation is used to rewrite a set of summed terms as a polynomial in the selected variable (if it is possible to do so). For example, the Expand function, operating on the x in $(x-3) \cdot (x-4)$, returned a polynomial. So we know that that expression can be written as a polynomial in x. The Collect function should also return that polynomial, and in the following example, we observe that it does:

$$(x-3) \cdot (x-4) \; \text{collect}, x \; \rightarrow \; x^2 - 7 \cdot x + 12$$

What is the difference between the Collect and Expand operations if they both return a polynomial? The Expand operation evaluates all powers and products of sums in the selected expression. For the expression $(x-3) \cdot (x-4)$, the result was a polynomial, but this is not always true. The Collect operation attempts to return a polynomial in the selected variable. For example, using the Expand operation on the result of a sum of terms with a common denominator yields a different result from that obtained using the Collect operation on the same sum, as is shown in these examples:

$$\frac{(a \cdot x - 4 \cdot a + b \cdot x - 3 \cdot b)}{((x-3) + (x-4))} \; \text{expand}, x \rightarrow \frac{1}{((x-3) \cdot (x-4))} \cdot a \cdot x$$

$$- \frac{4}{((x-3) \cdot (x-4))} \cdot a + \frac{1}{((x-3) \cdot (x-4))} \cdot b \cdot x - \frac{3}{((x-3) \cdot (x-4))} \cdot b$$

$$\frac{(a \cdot x - 4 \cdot a + b \cdot x - 3 \cdot b)}{((x-3) + (x-4))} \; \text{collect}, x \rightarrow \frac{((a+b) \cdot x - 4 \cdot a - 3 \cdot b))}{((x-3) \cdot (x-4))}$$

Partial-Fraction Expansion

A partial-fraction expansion on a variable is a method for expanding a complex expression into a sum of (we hope) simpler expressions with denominators containing only linear and quadratic terms and no functions of the variable in the numerator. As a first example, a partial-fraction expansion, based upon x, on

$$\frac{(a \cdot (x-4) + b \cdot (x-3))}{((x-3) \cdot (x-4))}$$

gives back the original function,

$$\frac{a}{(x-3)} + \frac{b}{(x-4)}$$

This was accomplished by selecting an x (any of them) and then choosing Symbolics/Variable/Convert to Partial Fraction from the Symbolics menu.

A slightly more complex example is to take the ratio of two polynomials and expand it by using partial fractions. The process is the same as described in the preceding example: Select the variable upon which you want to expand (z in this example), and then choose Symbolics/Variable/Convert to Partial Fraction from the Symbolics menu. The result will look like this:

$$\frac{(9 \cdot z^2 - 40 \cdot z + 23)}{(z^3 - 7 \cdot z^2 + 7 \cdot z + 15)}$$

$$\frac{2}{(z-3)} + \frac{3}{(z+1)} + \frac{4}{(z-5)}$$

An expanded example involving two variables illustrates how the partial fraction expansion depends on the variable selected. The starting expression now involves both z and y:

$$\frac{(7 \cdot z^3 + 19 \cdot z^2 - 43 \cdot z^2 \cdot y - 100 \cdot z \cdot y + 9 \cdot z - 33 \cdot y + 60 \cdot y^2 \cdot z + 105 \cdot y^2)}{(z^4 + 4 \cdot z^3 - 8 \cdot z^3 \cdot y - 32 \cdot z^2 \cdot y + 3 \cdot z^2 - 24 \cdot z \cdot y + 15 \cdot y^2 \cdot z^2 + 60 \cdot y^2 \cdot z + 45 \cdot y^2)}$$

A partial-fraction expansion on z requires that each denominator be linear or quadratic in z and precludes any function of z from every numerator. The result is

$$\frac{5}{(2 \cdot (z+3))} + \frac{3}{(2 \cdot (z+1))} - \frac{2}{(-z+5 \cdot y)} - \frac{1}{(-z+3 \cdot y)}$$

On the other hand, a partial-fraction expansion on y precludes functions of y in the numerators, but allows functions of z. The result of a partial-fraction expansion on y is

$$\frac{(4 \cdot z + 7)}{(z^2 + 4 \cdot z + 3)} - \frac{2}{(-z+5 \cdot y)} - \frac{1}{(-z+3 \cdot y)}$$

7.4 POLYNOMIAL COEFFICIENTS

If you have an expression that can be written as a polynomial, Mathcad will return a vector containing the coefficients of the polynomial. This operation is available from the Symbolics menu using Symbolics/Polynomial Coefficients.

The coefficients of the polynomial $x^2 - 7x + 12$ are pretty obvious. If you select any x in the expression, say,

$$\boxed{x^2 - 7x| + 12}$$

and then ask for the vector of polynomial coefficients as Symbolics/Polynomial Coefficients, the coefficients are returned as a vector:

$$x^2 - 7 \cdot x + 12$$

$$\begin{bmatrix} 12 \\ -7 \\ 1 \end{bmatrix}$$

Note: While Mathcad's symbolic math functions typically display the higher powers first (i.e., the x^2 before the $7x$), the first polynomial coefficient in the returned vector is the constant, 12.

When your expression is written as a polynomial, the coefficients are apparent. However Mathcad's symbolic processor can return the polynomial coefficients of expressions that are not displayed in standard polynomial form. For example, we know that $x^2 - 7x + 12$ is algebraically equivalent to $(x-3) \cdot (x-4)$. Using Symbolics/Polynomial Coefficients on $(x-3) \cdot (x-4)$ returns the same polynomial coefficients as in the standard-form case:

$$(x-3) \cdot (x-4)$$

$$\begin{bmatrix} 12 \\ -7 \\ 1 \end{bmatrix}$$

7.5 SYMBOLIC MATRIX MATH

Mathcad provides symbolic matrix operations from either the Symbolics menu or the Symbolic Keyword Toolbar. These are very straightforward operations and only summarized in this section.

Tranpose

To transpose a matrix, select the entire matrix and then choose Symbolics/Matrices/Transpose. The original and transposed matrices are, respectively,

$$\begin{bmatrix} 1 & 1 \\ 2 & 8 \\ 3 & 27 \\ 4 & 64 \\ 5 & 125 \end{bmatrix}$$

and

$$\begin{bmatrix} 1 & 2 & 3 & 4 & 5 \\ 1 & 8 & 27 & 64 & 125 \end{bmatrix}$$

Inverse

To invert a matrix symbolically, select the entire matrix and then choose Symbolics/Matrices/Invert. The original and inverted matrices are, respectively,

$$\begin{bmatrix} 2 & 3 & 5 \\ 7 & 2 & 4 \\ 8 & 11 & 6 \end{bmatrix}$$

and

$$\begin{bmatrix} \dfrac{-32}{211} & \dfrac{37}{211} & \dfrac{2}{211} \\ \dfrac{-10}{211} & \dfrac{-28}{211} & \dfrac{27}{211} \\ \dfrac{61}{211} & \dfrac{2}{211} & \dfrac{-17}{211} \end{bmatrix}$$

If you want to see the inverted matrix represented as values rather than fractions, select the entire inverted matrix, then choose Symbolics/Evaluate/Floating Point..., and, finally, enter the number of digits to display. Twenty digits is the default, but that is often excessive. Here, the result is displayed with a floating-point precision of four digits:

$$\begin{bmatrix} -.1517 & .1754 & .009479 \\ -.04739 & -.1327 & .128 \\ .2891 & .009479 & -.08057 \end{bmatrix}$$

Determinant

Mathcad will also symbolically calculate the determinant of a matrix. To do so, select the entire matrix and then choose Symbolics/Matrices/Determinant. In this example, the determinant is found to be 211:

$$\begin{bmatrix} 2 & 3 & 5 \\ 7 & 2 & 4 \\ 8 & 11 & 6 \end{bmatrix}$$
211

7.6 SYMBOLIC INTEGRATION

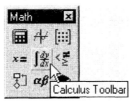

Mathcad provides a number of ways to integrate and differentiate functions. In this chapter, we focus on symbolic, or analytic integration, leaving numerical integration for the next chapter. The integration and differentiation operators are found on the *Calculus Toolbar*, which is available from the Math Toolbar.

In this section, we use a simple polynomial as a demonstration function and the same polynomial throughout to allow us to compare the various integration and differentiation methods, except where multiple variables are required to demonstrate multivariate integration. The polynomial is $12+3x-4x^2$, and a multivariable function, the formula for the volume of a cylinder, $\iiint r\, dr\, d\theta\, dl$ (with $0 \leq r \leq R$, $0 \leq \theta \leq 2\pi$, $0 \leq 1 \leq L$), is used to demonstrate multivariable integration and differentiation.

Indefinite Integrals

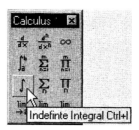

7.6.1 Symbolic Integration Using the Indefinite Integral Operator

The most straightforward of Mathcad's integration methods uses the *indefinite integral operator* from the Calculus Toolbar. When you click on the button for the indefinite integral operator, the operator is placed on your worksheet with two empty placeholders, the first for the function to be integrated and the second for the integration variable:

$$\int \blacksquare d\blacksquare$$

To integrate the sample polynomial, enter the polynomial in the first placeholder and an x in the second placeholder:

$$\int (12+3\cdot x-4\cdot x^2)dx$$

Complete the integration using the "evaluate symbolically" symbol, the $\rightarrow$. This symbol is entered by pressing [Ctrl-.] (Hold down the Control key while pressing the period key.) The result is

$$\int (12+3\cdot x-4\cdot x^2)dx \rightarrow 12\cdot x+\frac{3}{2}\cdot x^2-\frac{4}{3}\cdot x^3$$

Note that the integration variable is a dummy variable. You can put any variable you want in that placeholder. However, if you write the polynomial as a function of x and integrate with respect to y, Mathcad will perform the integration, but the result will not be very interesting:

$$\int (12+3\cdot x-4\cdot x^2)dx \rightarrow (12+3\cdot x-4\cdot x^2)\cdot y$$

Alternatively, you can select the entire expression and then choose Simplify from the Symbolics menu:

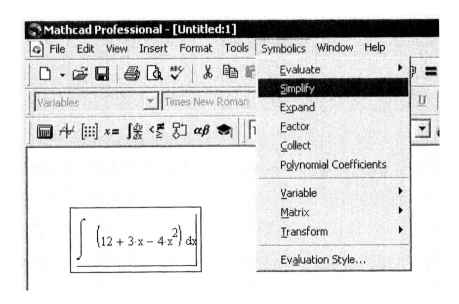

By default, the result will be placed below the integral operator:

$$\int (12 + 3 \cdot x - 4 \cdot x^2) dx$$

$$12 \cdot x + \frac{3}{2} \cdot x^2 - \frac{4}{3} \cdot x^3$$

You can modify the placement of the result by using the Symbolics/Evaluation style menu options.

Notice that Mathcad does not add the constant of integration you might expect when integrating without limits. Mathcad shows the functional form of the integrated expression, but you have to add your own integration constant if you want to evaluate the result. The integrated result is an editable equation region, so adding the constant is no problem:

$$12 \cdot x + \frac{3}{2} \cdot x^2 - \frac{4}{3} \cdot x^3 + C$$

7.6.2 Symbolic Integration with Multiple Variables

You can use the indefinite integration operator to integrate over multiple variables. Simply use one indefinite operator symbol for each integration variable. The second integration operator goes in the first operator's function placeholder, and so on. To integrate over three variables say, r, θ, and l, we'll use three integration operators:

$$\int \int \int \blacksquare d\blacksquare \ d\blacksquare \ d\blacksquare$$

Note that there is now only one function placeholder (before the first d), but there are three integration variable placeholders. In the computation of the volume of a cylinder, the function placeholder contains only an r:

$$\int \int \int r \ d\blacksquare d\blacksquare d\blacksquare$$

The integration variables are r, θ, and l:

$$\int\int\int r\ dr d\theta\ dl$$

You instruct Mathcad to evaluate the integral with the → operator:

$$\int\int\int r\ dr\ d\theta\ dl \rightarrow \frac{1}{2}\cdot r^2 \cdot \theta \cdot l$$

Perhaps this is not quite the result you anticipated. The expression was integrated as an indefinite integral. The π that you normally see in the formula for the volume of a cylinder comes from the integration limits on θ. This will become apparent later, when we integrate this function with limits (using the definite integral operator).

7.6.3 Symbolic Integration Using the Symbolics Menu

Symbolic integration using the Symbolics menu is an alternative method for integrating with respect to a *single variable*. Symbolic integration does not use the indefinite integral operator at all. Instead, you enter your function—for instance,

$$12+3\cdot x-4\cdot x^2$$

and then select the variable of integration, one of the x's in this example:

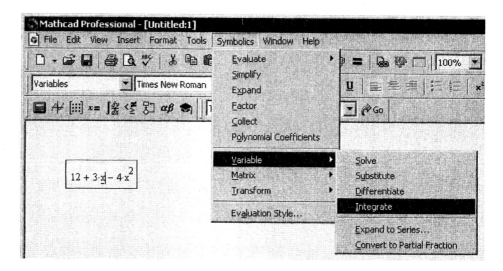

To perform the integration, use the Symbolics menu (Symbolics/Variable/Integrate) to obtain

$$12+3\cdot x-4\cdot x^2$$

$$12\cdot x+\frac{3}{2}\cdot x^2-\frac{4}{3}\cdot x^3$$

The result (again, without the constant of integration) is placed below the original function. This approach is very handy, but it is useful only for indefinite integrals with a single integration variable.

PRACTICE!

Use Mathcad's indefinite integral operator to evaluate the following integrals (starting with the obvious):

a. $\int x\ dx$.

b. $\int (a + bx + cx^2)dx$.

c. $\int\int 2x^2 y\, dx\, dy$.

d. $\int \dfrac{1}{x} dx$.

e. $\int e^{-ax} dx$.

f. $\int \cos(2x)dx$.

Definite Integrals

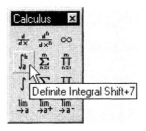

Mathcad evaluates definite integrals using the *definite integral operator* from the Calculus Toolbar. Definite integrals can be evaluated in a number of ways:

- symbolic evaluation with variable limits
- symbolic evaluation with numeric limits
- symbolic evaluation with mixed (variable and numeric) limits
- numerical evaluation (requires numeric limits)

7.6.4 Definite Integrals: Symbolic Evaluation with Variable Limits

The definite integral operator comes with four placeholders—the function and integration variable placeholders and two limit placeholders:

$$\int_\blacksquare^\blacksquare \blacksquare\, d\blacksquare$$

To evaluate the polynomial integrated from $x = A$ to $x = B$, simply include the limits in the appropriate placeholders:

$$\int_A^B (12 + 3 \cdot x - 4 \cdot x^2) dx$$

You instruct Mathcad to evaluate the integral either by using the symbolic evaluation symbol ($\rightarrow$) or by choosing Simplify from the Symbolics menu. In either case, the result is

$$\int_A^B (12 + 3 \cdot x - 4 \cdot x^2) dx \rightarrow 12 \cdot B - \dfrac{4}{3} \cdot B^3 + \dfrac{3}{2} \cdot B^2 - 12 \cdot A + \dfrac{4}{3} \cdot A^3 - \dfrac{3}{2} \cdot A^2$$

7.6.5 Definite Integrals: Symbolic Evaluation with Numeric Limits

To evaluate an integral with numeric limits, you can place the numbers in the limit placeholders, as, for example, in

$$\int_{-1}^{2} (12 + 3 \cdot x - 4 \cdot x^2) dx \rightarrow \frac{57}{2}$$

Or you can assign values to variables before the integration and use the variables as limits, as in

$$A := -1 \quad B := 2$$

$$\int_{A}^{B} (12 + 3 \cdot x - 4 \cdot x^2) dx \rightarrow \frac{57}{2}$$

7.6.6 Definite Integrals: Symbolic Evaluation with Mixed Limits

It is fairly common to have a numeric value for one limit and to want to integrate from that known value to an arbitrary (variable) limit. Mathcad handles this type of integration as well:

$$\int_{-1}^{C} (12 + 3 \cdot x - 4 \cdot x^2) dx \rightarrow 12 \cdot C - \frac{4}{3} \cdot C^3 + \frac{3}{2} \cdot C^2 + \frac{55}{6}$$

In the preceding example, the known limit was evaluated and generated the 55/6 in the result. The unknown limit was evaluated in terms of the variable C. The result can then be evaluated for any value of C.

Note: In evaluating an integral from a known limit to a variable limit, it is common to use the integration variable as the variable limit as well. Thus, for the preceding example, we would have

$$\int_{-1}^{x} (12 + 3 \cdot x - 4 \cdot x^2) dx$$

Mathcad, however will not evaluate this expression. For Mathcad, the integration variable (the x in dx) is a dummy variable, but the limits are not. In Mathcad, you cannot use the same symbol to represent both a dummy variable and a limit variable in a single equation.

7.6.7 Mixed Limits with Multiple Integration Variables

To obtain a formula for the volume of any cylinder, we would integrate the cylinder function over the variable r from 0 to some arbitrary radius R, over θ from 0 to 2π (for a completely round cylinder), and over l from 0 to an arbitrary length L. The result is the common expression for the volume of a cylinder:

$$\int_{0}^{L} \int_{0}^{2 \cdot \pi} \int_{0}^{R} r \, dr \, d\theta \, dl \rightarrow R^2 \cdot \pi \cdot L$$

Note the order of the integration symbols and the integration variables. Mathcad uses the limits on the inside integration symbol (0 to R) with the inside integration variable (dr), the limits on the middle integration operator (0 to 2π) with the middle integration variable (dθ), and so forth.

7.6.8 Definite Integrals: Numerical Evaluation

Numerical evaluation of an integral does not really fit in this chapter on symbolic math, but it is the final way that Mathcad can evaluate an integral. To request a numerical evaluation, use the equal sign instead of the $\rightarrow$ symbol:

$$\int_{-1}^{2}(12+3\cdot x-4\cdot x^2)dx = 28.5|$$

Normally, you would see the result to 20 decimal places (by default), but this result is precise, at 28.5. Note that the result includes a units placeholder. Units can be used with numerical evaluation (and with the limits on symbolic integration as well).

PRACTICE!

Use Mathcad's definite integral operator to evaluate these expressions:

a. $\int_{0}^{4} x\,dx.$

b. $\int_{1}^{3} (ax+b)\,dx.$

c. $\int_{-3}^{0} \frac{1}{3}\,dx.$

d. $\int_{0}^{\pi}\int_{0}^{2\,cm} r\,dr\,d\theta.$

PRACTICE!

Find the area under the following curves in the range $x = 1$ and $x = 5$ (Part a has been completed as an example):

a. $3 + 1.5x - 0.25x^2$

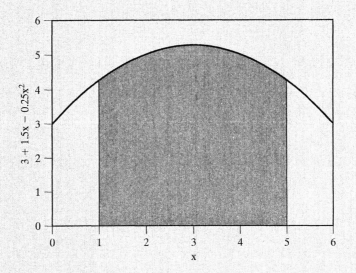

Area $= \int_{1}^{5} 3 + 1.5x - 0.25x^2\,dx$

$= 19.667$

b. $0.2 + 1.7x^3$

c. $\sin\left[\dfrac{x}{2}\right]$

Find the area between the following curves in the range $x = 1$ and $x = 5$ (Part a has been completed as an example):

a. $3 + 1.5x - 0.25x^2$ and $0 + 1.5x - 0.25x^2$

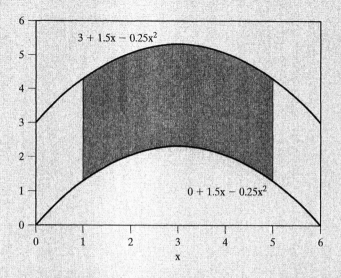

$$\text{Area} = \int_1^5 (3 + 1.5x - 0.25x^2) - (0 + 1.5x - 0.25x^2) dx$$
$$= \int_1^5 3 dx$$
$$= 12$$

b. $3 + 1.5x - 0.25x^2$ and $0.1x^2$

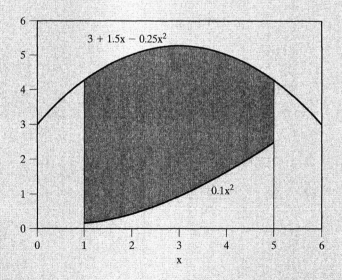

APPLICATIONS: ENERGY REQUIRED TO WARM A GAS

Warming up a gas is a pretty common thing to do, partly because we like to live in warm buildings and partly because we tend to burn things to warm those buildings. Combustion always warms up its products. It is common to need to know how much energy is required to warm a gas.

The amount of energy needed to warm a gas depends on the amount, heat capacity, and temperature change of the gas. Also, heat capacities of gases are strong functions of temperature, so the relationship between heat capacity and temperature must be taken into account. All of this is included in the equation

$$\Delta H = n \int_{T_1}^{T_2} C_p \, dT$$

where

ΔH is the change in enthalpy of the gas, which is equal to the amount of energy required to warm the gas if all of the energy added to the gas is used to warm it (i.e., if the energy is not used to make the gas move faster, etc.),

n is the number of moles of gas present (say, 3 moles),

C_p is the heat capacity of the gas at constant pressure,

T_1 is the initial temperature of the gas (say, 25°C), and

T_2 is the final temperature of the gas (say, 400°C).

Since heat capacities change with temperature, the relationship between heat capacity and temperature is often given as an equation. For example, for CO_2, the heat capacity is related to the temperature by the expression[1]

$$C_p = 36.11 + 4.233 \cdot 10^{-2}T - 2.887 \cdot 10^{-5}T^2 + 7.464 \cdot 10^{-9}T^3 \quad (\text{J/mole °C}),$$

and the equation is valid for temperatures between 0 and 1500°C.

We can use Mathcad to integrate this expression and determine the amount of energy required to warm 3 moles of CO_2 from 25 to 400°C:

$$n := 3$$

$$\Delta H := n \cdot \int_{25}^{400} (36.11 + 4.233 \cdot 10^{-2} \cdot T - 2.887 \cdot 10^{-5} \cdot T^2 + 7.464 \cdot 10^{-9} \cdot T^3) dT$$

$$\Delta H = 4.904 \cdot 10^4$$

The energy required is 49 kJ.

Note that this problem was worked without units for two reasons:

1. The units on the various terms in the heat capacity equation are complicated.
2. Mathcad doesn't have °C as a built-in unit, and the T's in this heat capacity equation must be in °C.

[1] From *Elementary Principles of Chemical Processes* by R. M. Felder and R. W. Rousseau, 2d ed., Wiley, New York (1986).

7.7 SYMBOLIC DIFFERENTIATION

You can evaluate derivatives with respect to one or more variables using the *Derivative* or N^{th} *Derivative* buttons on the Calculus Toolbar. For a first derivative with respect to a single variable, you can also use Variable/Derivative from the Symbolics menu.

7.7.1 First Derivative with Respect to one Variable

When you click on the Derivative button on the Calculus Toolbar, the derivative operator is placed on the worksheet:

$$\frac{d}{d\blacksquare}\blacksquare$$

The operator contains two placeholders—one for the function, the other for the differentiation variable. To take the derivative of the sample polynomial with respect to the variable x, the polynomial and the variable are inserted into their respective placeholders:

$$\frac{d}{dx}(12+3\cdot x-4\cdot x^2)$$

You tell Mathcad to evaluate the derivative either by using the → symbol, producing

$$\frac{d}{dx}(12+3\cdot x-4\cdot x^2) \rightarrow 3-8\cdot x$$

or by selecting the entire expression and then choosing Simplify from the Symbolics menu, resulting in

$$\frac{d}{dx}(12+3\cdot x-4\cdot x^2)$$

$$3-8\cdot x$$

Note that Mathcad does not have a "Derivative evaluated at" operator. To evaluate the result at a particular value of x, simply give x a value before performing the differentiation:

$$x := -1$$

$$\frac{d}{dx}(12+3\cdot x-4\cdot x^2) \rightarrow 11$$

As an alternative to using the Derivative operator from the Calculus Toolbar, you can enter your function and select one variable (one of the x's in the polynomial example):

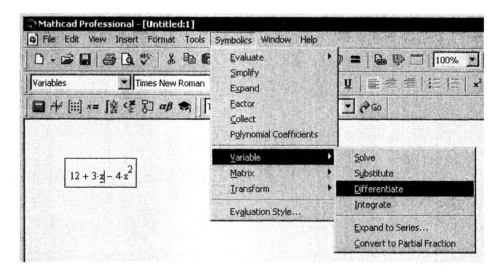

Then you differentiate the function with respect to the selected variable using Symbolics/Variable/Differentiate from the Symbolics menu. The result will be

$$12+3\cdot x-4\cdot x^2$$

$$3-8\cdot x$$

This method works only for evaluating a first derivative with respect to a single variable.

7.7.2 Higher Derivatives with Respect to a Single Variable

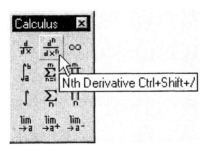

For higher derivatives, use the Nth Derivative operator from the Calculus Toolbox. This operator comes with four placeholders, but you can use only three:

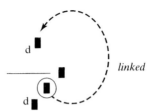

When you add the power to the right placeholder in the denominator, the same power will appear in the numerator of the derivative operator. You cannot type directly into the placeholder in the numerator. In this example, we'll use a power of 2 to take the second derivative of the sample polynomial:

$$\frac{d^2}{d\blacksquare^2}\blacksquare$$

The two remaining placeholders are for the function and the differentiation variable:

$$\frac{d^2}{dx^2}(12+3\cdot x-4\cdot x^2) \rightarrow -8$$

7.7.3 Differentiation with Respect to Multiple Variables

Use multiple derivative operators to evaluate derivatives with respect to multiple variables. For example, the indefinite integral of r with respect to r, θ, and 1 yields this result:

$$\int\int\int r \; dr \; d\theta \; d1 \rightarrow \frac{1}{2}\cdot r^2 \cdot \theta \cdot 1$$

If we take the result and differentiate it with respect to r, θ, and 1, we should get the original function back:

$$\frac{d}{d1}\frac{d}{d\theta}\frac{d}{dr}\left(\frac{1}{2}\cdot r^2 \cdot \theta \cdot 1\right) \rightarrow r$$

The original function is simply r, all by itself.

As a more interesting example, consider the ideal-gas law, and take the derivative of pressure with respect to temperature and volume:

$$P = \frac{n \cdot R \cdot T}{V}$$

$$\frac{d}{dT}\frac{d}{dV}\left(\frac{n \cdot R \cdot T}{V}\right) \rightarrow -n \cdot \frac{R}{V^2}$$

PRACTICE!

Try using Mathcad to evaluate these derivatives:

a. $\dfrac{d}{dx} x^2$.

b. $\dfrac{d}{dx}(3x^2 + 4x)$.

c. $\dfrac{d^2}{dx^2} x^3$.

d. $\dfrac{d}{dx}\dfrac{d}{dy}(3x^2 + 4xy + 2y^2)$.

e. $\dfrac{d}{dx} \ln(ax^2)$.

f. $\dfrac{d}{dx} \cos(2x)$.

APPLICATION: ANALYZING STRESS–STRAIN DIAGRAMS

A fairly standard test for new materials is the tensile test. In essence, a sample of the material is very carefully prepared and then slowly pulled apart. The stress on the sample and the elongation of the sample are recorded throughout the test. We will consider data from a composite material in this example, but to allow for comparison, we first consider a tensile test on a metal sample. The stress–strain diagram obtained from testing is as follows:

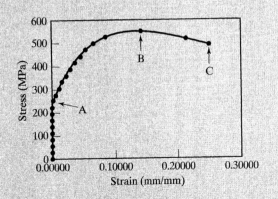

Note the units on strain in the foregoing diagram: mm/mm. This is millimeters of elongation divided by the original length of the sample. The test begins with no stress and no strain. As the pulling begins, the metal starts to stretch. From the origin of the graph to point A, the stretching is reversible: If the pulling pressure were released, the metal would return to its original size. Beyond point A, some of the stretching is irreversible. Point B is called the material's *ultimate stress*—the highest stress that the material can withstand without breaking. Beyond point B, the stress actually goes down as the sample pulls itself apart under the applied stress. At point C, the sample breaks.

The stress–strain curves for *composite materials* have a different shape because of the way the materials respond to stress. There are many different composite materials with vastly different mechanical properties, so their stress–strain curves could be very different, but the curve shown in the following figure illustrates some interesting features:

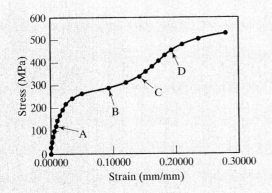

The big difference between this curve and the previous one is the presence of a second hump. Between the origin and point B, the curve looks a lot like an ordinary stress–strain curve, and then the second hump starts appearing at about point C. The explanation for this behavior is that the first hump represents mostly the matrix (surrounding the fibers) responding to the stress. The matrix then cracks and separates from the fibers, and the stress is transferred to the fibers between points B and C. From point C on, you are seeing the stress–strain response of the fibers.

There are a couple of analyses we can perform on these data:

a. Estimate Young's modulus for the matrix and the fibers.
b. Calculate the work done on the sample during the test.

YOUNG'S MODULUS

Young's modulus is the proportionality factor relating stress and strain in the linear sections of the graph between the origin and point A and (sometimes visible) between points C and D. Because the material is a composite, neither of the values we will calculate truly represents Young's modulus for the pure materials, but they will help quantify how this composite material behaves under stress.

The linear region near the origin includes approximately the first four or five data points. Young's modulus for the matrix can be calculated from the change in stress and the measured change in strain:

$$Y := \frac{Stress_3 - Stress_0}{Strain_3 - Strain_0}$$

$$Y = 5.929 \cdot 10^4 \quad \ll MPa$$

Similarly, Young's modulus relating stress to strain in the region between C and D includes points 15–18:

$$Y_{fiber} := \frac{Stress_{18} - Stress_{15}}{Strain_{18} - Strain_{15}}$$

$$Y_{fiber} = 1.012 \cdot 10^4 \quad \ll Mpa$$

WORK

A little reshaping can turn a stress–strain diagram into a force–displacement diagram. The area under a force–displacement diagram is the work done on the sample.

Stress is the force per unit cross-sectional area of the sample. If the sample tested is L = 10 mm by W = 10 mm, the area and force on the sample can be computed as

$$A := L \cdot W$$

$$F := Stress \cdot A$$

To obtain a displacement, x, we need to multiply the strain by the original sample length (or height; the samples are usually vertical when tested). If H = 10 mm as well then

$$x := Strain \cdot H$$

We can now replot the stress–strain diagram as a force–displacement graph. The result will look like this:

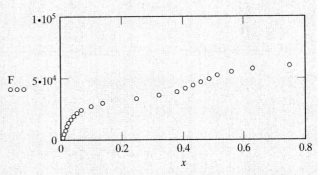

The area under the graph is the work, but in order to use Mathcad's integration operator, we need a function relating force to displacement, not data points. In Chapter 5 we saw how to fit a polynomial to data. That's what we need here. (*Note:* In the next chapter, integration methods using the data points themselves will be covered.)

Accordingly,

$$f(x) := \begin{bmatrix} x \\ x^2 \\ x^3 \\ x^4 \\ x^5 \end{bmatrix} \quad b := \text{linfit}(x, F, f)$$

$$\vec{F_p} := \overline{(b_0 \cdot x + b_1 \cdot x^2 + b_2 \cdot x^3 + b_3 \cdot x^4 + b_4 \cdot x^5)}$$

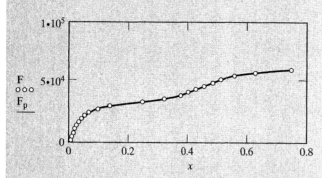

From the preceding graph, it looks like the fifth-order polynomial fits the data nicely. But F_p is a vector of values; we still need a *function*. Hence,

$$F_{\text{func}}(X) := b_0 \cdot x + b_1 \cdot x^2 + b_2 \cdot x^3 + b_3 \cdot x^4 + b_4 \cdot x^5$$

This function can be integrated using the definite integration operator from the Calculus Toolbox. As the graph shows, the upper limit on x is almost 0.8. The actual value can be found by using the max() function on the x vector:

$$\text{Work} := \int_0^{\max(x)} F_{\text{func}}(x) \, dx$$

$$\text{Work} = 3.069 \cdot 10^4$$

That's the work, but what are the units? F was determined from stress (MPa) and area (mm²), and x came from strain (mm/mm) and length (mm). This work has units of MPa · (mm³). We'll convert them:

$$\text{Work} := \text{Work} \cdot \text{MPa} \cdot \text{mm}^3$$

$$\text{Work} = 30.695 \cdot N$$

Note: The integration operator can handle units, but the linfit() function does not. That's why this problem was worked without units.

SUMMARY

In this chapter, we looked at Mathcad's symbolic math capabilities—its ability to work directly with mathematical expressions, rather than numerical results. We saw that Mathcad's symbolic math features are housed in two areas and are used in slightly different ways. For example, you can solve for a variable in an equation from either the Symbolics menu or the Symbolic Keyword Toolbar. With the latter, there is a "live operator," so that if you make changes to the worksheet, the solve operation will be automatically recalculated, and you specify the variable you want to solve for as part of the operation. In order to solve for a variable using the Symbolics menu, you first select the variable of interest and then use Symbolics/Variable/Solve from the menu. The result is placed on the worksheet, but it is not a live operator, so the result will not be automatically recalculated if the worksheet changes. Both approaches are useful.

You also saw that Mathcad can replace (substitute) every occurrence of a variable with another mathematical expression and can factor common terms out of complex expressions. You can use symbolic math to find a common denominator using the simplify operation. Mathcad can manipulate expressions in a number of other ways as well.

Finally, you learned that Mathcad can do matrix math operations symbolically, as well as integrate and differentiate expressions.

MATHCAD SUMMARY

Operations under the Symbolics Menu:

Before using the Symbolics menu, you generally need to select the part of an expression you want to operate on. For example, you need to select a variable in an expression before using any of the Symbolics/ Variable operations.

Symbolics/Simplify	Evaluates common math operations (e.g., square root) to try to simplify an expression.
Symbolics/Expand	Multiplies out powers and polynomials, expands numerators of fractions.
Symbolics/Factor	Reverses the expand operation: simplifies polynomials, pulls a common expression out of multiple terms.
Symbolics/Collect	Tries to rewrite a set of summed terms as a polynomial.
Symbolics/Polynomial Coefficients	If the selected expression can be written as a polynomial, returns the polynomial coefficients.
Symbolics/Variable/Solve	Solves an expression for the selected variable.
Symbolics/Variable/Substitute	Replaces each occurrence of the selected variable with the expression in the Windows clipboard.
Symbolics/Variable/Differentiate	Returns the first derivative of the expression with respect to the selected variable.
Symbolics/Variable/Integrate	Integrates the expression with respect to the selected variable—does not add a constant of integration.
Symbolics/Variable/Convert to Partial Fractions	Expands an expression into a sum of expressions with denominators containing only linear and quadratic terms and with no functions of the selected variable in the numerator.
Symbolics/Matrix/Transpose	Interchanges rows and columns in the matrix.
Symbolics/Matrix/Invert	Inverts the matrix using symbolic math operations, rather than decimal numbers—this preserves accuracy by avoiding round-off errors, but the resulting fractions can get unwieldy.
Symbolics/Matrix/Determinant	Calculates the determinant of a matrix using symbolic math operations.
Symbolics/Evaluation Style...	Opens a dialog box that allows you to change the way the results of the calculations are presented.

Operations Using the Symbolic Keyword Toolbox:

With the Symbolic Keyword Toolbox, the symbolic evaluation operator → is used and any variables or expressions that must be specified in order to perform the operation are entered into placeholders as needed. You do not need to select a variable or a portion of the expression before performing the symbolic operation.

solve	Solves the expression for the specified variable.
simplify	Evaluates common math operations to try to simplify an expression.
substitute	Replaces each occurrence of the specified variable with the specified expression.

factor	Pulls the specified variable out of all terms in an expression.
expand	Evaluates powers and polynomials involving the specified variable; expands numerators of fractions involving the variable.
coeffs	Returns the polynomial coefficients, if there are any, for the specified variable.
collect	Tries to rewrite summed terms as a polynomial in the specified variable.
parfrac	Expands an expression into a sum of expressions, with denominators containing only linear and quadratic terms and with no functions of the selected variable in the numerator.
$M^T \rightarrow$	Transposes the matrix (interchanges rows and columns in the matrix).
$M^{-1} \rightarrow$	Inverts the matrix using symbolic math operations.
$|M| \rightarrow$	Calculates the determinant of a matrix using symbolic math operations.

Operations Using the Calculus Palette:

Indefinite Integral	Integrates the expression in the function placeholder with respect to the variable in the integration variable placeholder. If the symbolic evaluation operator $\rightarrow$ is used after the integral, the integral is evaluated symbolically. Mathcad does not add a constant of integration. If an equal sign is used after the integral, the integral is evaluated numerically.
Definite Integral	Integrates the expression in the function placeholder with respect to the variable in the integration variable placeholder, using the specified limits. (The dummy integration variable cannot be used in a limit.) If the symbolic evaluation operator $\rightarrow$ is used after the integral, the integral is evaluated symbolically. If an equal sign is used after the integral, the integral is evaluated numerically.
Derivative	Takes the derivative of the expression in the function placeholder with respect to the variable in the differentiation variable placeholder. If the symbolic evaluation operator $\rightarrow$ is used, the derivative is evaluated symbolically. If an equal sign is used, the derivative is evaluated numerically. To evaluate the derivative at a particular value, assign the value to the differentiation variable before taking the derivative.
Nth Derivative	Takes second- and higher-order derivatives.

KEY TERMS

collect terms
definite integral
derivative
differentiation
expand

factor
indefinite integral
limits of integration
live symbolic operator ($\rightarrow$)
partial-fraction expansion

polynomial coefficients
simplify
substitute
symbolic math

Problems

1. **SOLVING FOR A VARIABLE IN AN EXPRESSION**
   The equation
   $$A = 2 \cdot p \cdot r^2 + 2 \cdot p \cdot r \cdot L$$
   calculates the surface area of a cylinder. The equation is written using the symbolic equality, [CTRL-=].
   a. Use the `solve` operation on the Symbolic Keyword Toolbar to solve for the L in this equation.
   b. Use the expression returned by the `solve` operation to determine the cylinder length required to get 1 m² of surface area on a cylinder with a radius of 12 cm.

2. **POLYNOMIAL EQUATIONS: COEFFICIENTS AND ROOTS**
   The following equation clearly has three solutions (roots), at $x = 3$, $x = 1$, and $x = -2$:
   $$(x - 3) \cdot (x - 1) \cdot (x + 2) = 0$$
   a. Create a QuickPlot of $(x - 3) \cdot (x - 1) \cdot (x + 2)$ vs. x in the range from -3 to 3. Verify that the curve does cross the x-axis at $x = 3$, $x = 1$, and $x = -2$.
   b. Use the `solve` operation on x to have Mathcad find the roots:
   $$(x-3) \cdot (x-1) \cdot (x+2) \quad \text{solve,} \quad x \rightarrow$$
   c. Expand the function in x to see the expression written as a polynomial:
   $$(x-3) \cdot (x-1) \cdot (x-2) \quad \text{expand,} \quad x \rightarrow$$
   d. Obtain the polynomial coefficients as a vector using the `coeffs` operator from the Symbolic Math Toolbar:
   $$(x-3) \cdot (x-1) \cdot (x-2) \quad \text{coeffs,} \quad x \rightarrow$$

3. **SYMBOLIC INTEGRATION**
   Evaluate the following indefinite integrals symbolically:
   a. $\int \sin(x)\, dx$.
   b. $\int \ln(x)\, dx$.
   c. $\int [\sin(x)^2 + \cos(x)]\, dx$.
   d. $\int_{-3}^{0} \dfrac{x}{x + b}\, dx$.

Note: In part c, notice how Mathcad indicates the sine-squared term: sin(x)², not sin²(x).

4. **DEFINITE INTEGRALS**

   Evaluate the following definite integrals, using either symbolic or numeric evaluations:

   a. $\int_0^{\pi} \sin(\theta)\, d\theta.$

   b. $\int_0^{2\pi} \sin(\theta)\, d\theta.$

   c. $\int_{-3}^{0} \frac{x}{x-3}\, dx.$

   d. $\int_0^{\infty} e^{\frac{-t}{4}}\, dt.$

   Note: The numerical integrator cannot handle the infinite limit in part d; use symbolic integration. The infinity symbol is available on the Calculus Palette.

5. **INTEGRATING FOR THE AREA UNDER A CURVE**

   Integrate $\int y\, dx$ to find the area under the curves represented by these functions:

   a. $y = -x^2 + 16$ between $-2 \leq x \leq 4$
   b. $y = e^{\frac{-t}{4}}$ between $0 \leq t \leq \infty$
   c. $y = e^{\frac{-t}{4}}$ between $0 \leq t \leq 4$

6. **INTEGRATING FOR THE AREA BETWEEN CURVES**

   Find the area between the curves represented by the functions

   $$x = -x^2 + 16$$

   and

   $$y = -x^2 + 9$$

   over the range $-3 \leq x \leq 3$

7. **AREA OF A SECTOR**

   The technical term for a pie-slice-shaped piece of a circle is a *sector*. The area of a sector can be found by integrating over r and θ:

   $$A = \int_{\theta=0}^{\alpha} \int_{r=0}^{R} r\, dr\, d\theta$$

   a. Check the preceding equation by symbolically integrating over the entire circle ($\alpha = 2\pi$). Do you get the expected result, $A = \pi R^2$?

   b. Find the area of a 37° sector using Mathcad's numerical integration capability (i.e., use an equal sign rather than a symbolic evaluation operator ($\rightarrow$) to evaluate the integral.)

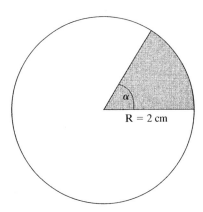

8. **DESIGNING AN IRRIGATION SYSTEM**

   Problem 2.15 introduced center-pivot irrigation systems and described these systems in a general fashion. An essential element of such systems is their ability to distribute water fairly evenly. Because the end of the pipe covers a lot more ground than the pipe near the center, water must be applied at a faster rate at the outside of the circle than near the center in order to apply the same number of gallons per square foot of ground.

   a. Use Mathcad's symbolic math capabilities to integrate the equation

   $$\text{Area}_{ring} = \int_0^{2\pi} \int_{R_i}^{R_o} r \, dr \, d\theta$$

   to obtain a formula for the area of a ring with inside radius R_i and outside radius R_o.

   b. At what rate must water be applied to provide one inch (depth) of water
      i. to the innermost ring: $R_i = 0$, $R_o = 20$ feet.
      ii. to the outermost ring: $R_i = 1300$ feet, $R_o = 1320$ feet.

   c. Write a Mathcad function that accepts the inside and outside radii and the desired water depth as inputs and returns the flow rate required to provide one inch of water. Use your function to determine the water flow rate for $R_i = 1000$ feet, $R_o = 1020$ feet, and depth = 2 inches.

9. **FINDING THE EQUATION FOR THE AREA OF AN ELLIPSE**

   The equation of an ellipse centered at the origin is

   $$\frac{x^2}{a^2} + \frac{y^2}{b^2} = 1$$

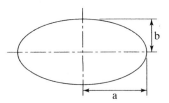

The area of the upper half of the ellipse can be determined by finding the area between the ellipse and the x-axis ($y = 0$):

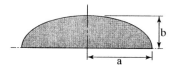

The total area of the ellipse is twice the area of the upper half.

a. Solve the equation of the ellipse for y. (You will obtain two solutions, since there are two y values on the ellipse at every x value.) Verify that the positive y values are returned by

$$y = \frac{b}{a} \cdot \sqrt{-(x^2) + a^2}$$

b. Substitute the expression for y from part a in the integral for the area of an ellipse (twice the area of the upper half of the ellipse), and solve the following equation for the area of an ellipse symbolically:

$$2 \cdot \int_{-a}^{a} y \, dx \rightarrow$$

10. **WORK REQUIRED TO STRETCH A SPRING**

    Hooke's law says that the force exerted against a spring as the spring is being stretched is proportional to the extended length. That is,

    $$F = kx.$$

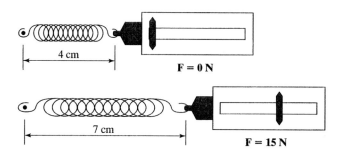

Work is defined as

$$W = \int F \, dx$$

From the accompanying figure, it can be seen that the length of the spring with no applied force is 4 cm. An applied force of 15 N extends the spring 3 cm to a total length of 7 cm. This information can be used to find the spring constant k. We write

$$15 \text{ N} = k \times 3 \text{ cm},$$
$$k = 5 \frac{\text{N}}{\text{cm}}$$

Calculate the work done on the spring as it was stretched from 4 to 7 cm (i.e., as the extended length went from 0 to 3 cm).

11. **WORK REQUIRED TO COMPRESS AN IDEAL GAS AT CONSTANT TEMPERATURE**

    One equation for work is

    $$W = \int P \, dV$$

    For an ideal gas, pressure and volume are related through the ideal gas law:

    $$PV = nRT$$

    a. Use Mathcad's ability to solve an expression for a variable to solve the ideal gas law for pressure P.
    b. Substitute the result for P from part a into the work integral.
    c. Determine the work required to compress 10 moles of an ideal gas at 400 K from a volume of 300 liters to a volume of 30 liters. Express your result in kJ. (Assume that cooling is provided to maintain the temperature at 400 K throughout the compression.)
    d. Does the calculated work represent work done on the system (the 10 moles of gas) by the surroundings (the outside world) or by the system on the surroundings?

12. **ENERGY REQUIRED TO WARM A GAS**

    At what rate must energy be added to a stream of methane to warm it from 20°C to 240°C? Data are as follows:

Methane flow rate:	20,000 mole/min (these are gram moles);
Heat capacity equation:[1]	$C_p = 34.31 + 5.469 \cdot 10^{-2} \, T + 0.3361 \cdot 10^{-5} \, T^2 - 11.00 \cdot 109 \, T^3$ J/mole(C).

13. **ANNULAR PIPING SYSTEMS**

    One pipe may be placed inside of another pipe, to carry two fluids, *creating an annulus*. One fluid flows in the center pipe, the other in the annular space between the two pipes. Annular flow arrangements are often used to allow heat transfer between the two fluids. For example, hot steam from a power plant might flow in the center pipe out to a distant laboratory to provide heat. In the building's heating system, the steam is condensed to keep the building warm. The liquid condensate is returned to the boiler in the annular space surrounding the steam pipe. In this way, the hot steam is somewhat insulated by the hot

---
[1]From *Elementary Principles of Chemical Processes* by R. M. Felder and R. W. Rousseau, 2d ed., Wiley, New York (1986).

condensate, and the condensate would be partially reheated by the steam as it returned to the power plant.

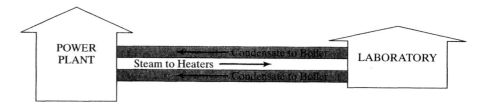

The annulus is the space between the two concentric pipes. The cross-sectional area of the annulus is an important design parameter used in sizing this type of piping system. The area of the annulus depends on the outside radius of the small pipe, r_o, and the inside radius of the large pipe, R_I.

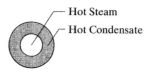

The area can be calculated in two ways:
- by subtracting the area of the small circle from the area of the large circle–that is,

$$A = \pi R_I^2 - \pi r_o^2$$

- by integrating the expression from r_O to R_I.

$$A = \int_{\theta=0}^{2\pi} \int_{r=r_o}^{R_I} r \, dr \, \theta$$

Using symbolic integration, demonstrate that the two methods give equivalent results.

14. YOUNG'S MODULUS

In the Mathcad Application analyzing stress–strain diagrams, values of Young's modulus for the matrix and the fiber were determined using algebra. An alternative approach would be to take the derivative of a function fit to the stress–strain data and evaluate the derivative in the linear regions of the stress–strain curve.

Let

$$f(x) := \begin{bmatrix} x \\ x^1 \\ x^3 \\ x^4 \\ x^5 \end{bmatrix} \quad b := \text{linfit (Strain, Stress, f)} \quad b = \begin{bmatrix} 5.951 \cdot 10^4 \\ -4.006 \cdot 10^6 \\ 1.195 \cdot 10^8 \\ -1.542 \cdot 10^9 \\ 7.184 \cdot 10^9 \end{bmatrix}$$

be a fifth-order polynomial fit to the stress–strain curves.

Use Mathcad's derivative operator to differentiate the polynomial and then evaluate the derivative at strain values of 0.001 and 0.048 to find the values of Young's modulus for the matrix and the fibers, respectively.

15. **CALCULATING WORK**

The stress–strain data for the metal sample mentioned in the Application box at the end of this chapter have been abridged and reproduced as follows:

STRAIN (MM/MM)	STRESS (MPA)
0.00000	0
0.00028	55
0.00055	110
0.00083	165
0.00110	221
0.00414	276
0.01324	331
0.02703	386
0.04193	441
0.06207	496
0.13793	552
0.20966	524
0.24828	496

a. Convert the following stress–strain diagram to a force–displacement graph (the sample size is 10 mm × 10 mm × 10 mm):

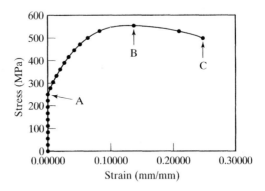

b. Fit a polynomial to the force–displacement data. (A fifth-order polynomial with no intercept works well.)

c. Integrate the polynomial to determine the work done on the sample during the tensile test.

8
Numerical Techniques

ENGINEERING IN MEDICINE

Medical science has developed dramatically in the past generation, and with the increased understanding of the inner workings of the human body in health and disease comes the ability to apply that knowledge to try to improve the human condition. Engineers from every discipline can play a significant role in this enterprise. Many bodily processes are traditional areas of engineering study, but usually with a slight twist or complicating factor. For example, most engineers are familiar with fluid flows in tubes, but blood is a non-Newtonian fluid in pulsatile flow in nonrigid tubes. Each of these factors complicates the study of blood flow in the human body, but the engineer's knowledge of steady, Newtonian flow in rigid tubes provides a good foundation for the study of the more complex system. Similar examples can be cited in other traditional fields of study, such as heat transfer (mechanisms for maintaining the body's temperature), mass transfer (carbon dioxide transfer from the blood in the lungs), reactor dynamics (controlled release of drugs into the body), mechanics (artificial limbs, improved athletic performance), and materials science (biologically inert materials for implants). In each of these areas engineers will work together, along with medical professionals, to expand our understanding of how our bodies work and to apply this knowledge to improve health care.

SECTIONS

8.1 Iterative Solutions
8.2 Numerical Integration
8.3 Numerical Differentiation

OBJECTIVES

After reading this chapter, you will

- know how to use Mathcad's powerful iterative solver
- be able to integrate by fitting data with a model equation, then integrating the equation
- be able to perform integration on a data set using numerical methods directly on the data values
- be able to differentiate by fitting data with a model equation, then differentiating the equation
- know how to use finite differences to calculate derivatives from data values

As only one example, designing an artificial limb requires a knowledge of anatomy and mechanics to design the right motions, a knowledge of physiology and circuit theory to understand the electrical signals in the body, and a knowledge of control systems to use the signals to control the limb. Where the circuitry, the prosthesis, and the body come in contact, materials must be carefully chosen to meet mechanical, electrical, and biological constraints. This is definitely a field for team players, and the opportunities for engineers to make significant contributions in the field of medicine have never been greater.

8.1 ITERATIVE SOLUTIONS

Equations (or systems of equations) that cannot be solved directly occur often in engineering. For example, to determine the flow rate in a pipe with a given pressure drop, the friction loss must be known. But the friction loss depends on the flow rate. Thus, in order to calculate the friction loss, you must know the flow rate. But to calculate the flow rate, you must know the friction loss. To solve this very common problem, it is necessary to guess either the flow rate or the friction loss. If you guess the friction loss, you would calculate the flow rate and then calculate the friction loss at that flow rate. If the calculated friction loss equals the guessed friction loss, the problem is solved. If not, guess again.

Mathcad provides a better way to solve problems like this: the *iterative solver*. As a simple example, consider finding the value of x that satisfies the equation

$$x^3 + 12x - 21 = 0.$$

The x^3 suggests that there will be three solutions, but there could be duplicate roots or imaginary roots. In this example, two of the roots are imaginary. We'll try to find the one real root.

Using the Worksheet for Trial-and-Error Calculations

You can simply do trial-and-error calculations in the Mathcad workspace. For example, if we try x = 0, the expression $x^3 + 12x - 21$ yields −21. Since −21 ≠ 0, the guessed value of x is incorrect. We might then try x = 1:

```
x := 1
x³+12·x-21 = -8
```

This result is closer to zero than the first attempt, so we're moving in the right direction, but the guessed x is still too small. Try x = 2:

```
x := 2
x³+12·x-21 = 11
```

The equation is still not satisfied, but this guess was too big. Try x = 1.5:

```
x :=  1.5
x³+12·x-21 = 0.375
```

We're getting close, but x = 1.5 is still a little high. With a few more tries, you'll find that a value of x = 1.48 comes very close to satisfying this equation.

Note: Iterative solutions always attempt to find values that "nearly" solve the equation. The process of choosing ever-closer guessed values could go on forever. To decide when the solution is "close enough" to stop the process, iterative methods test the calculated result against a preset *tolerance*. For Mathcad's iterative solver, the two sides of the equation are evaluated by using the computed value, and the difference between the two values is called the *error*. When the error falls below the preset tolerance, Mathcad stops iterating and presents the result. You can change the value of the tolerance from the default value of 0.001 to a larger number for less accurate solutions or (more likely) a smaller value for more accurate solutions. Very small tolerance values, such as 10^{-15}, may make it hard for Mathcad to find a solution because of computer round-off error.

Automating the Iterative Solution Process

Mathcad provides a better iterative solver than manual trial and error. Mathcad's approach uses an *iterative solve block*, which will allow you to solve multiple equations simultaneously. The solve block is bounded by two keywords: `given` and `find`. The equations between these two words will be included in the iterative search for a solution. Before the `given`, you must provide an initial guess for each variable in the solve block. The iterative solve block for the preceding cubic example, with an initial guess of x = 0, would look like this:

```
x :=   0
given
    x³ + 12·x - 21 = 0
x :=   find(x)
x = 1.4799
```

The equation between the `given` and `find` has been indented for readability. The line

```
x :=   find(x)
```

terminates the solve block and assigns the solution (returned by the `find()` function) to the variable x. Here the x variable was used to hold both the initial guess and the computed solution. This is common, but not necessary; you could assign the solution to any variable.

You can check the solution by using the computed value in the equation:

```
x :=   1.4799
x³ + 12·x - 21 = 0
```

You can adjust the displayed precision of any result by double-clicking on the displayed value and changing the number of displayed digits, but Mathcad's default is to drop trailing zeros. The zero on the right side of the equation is actually zero to at least three decimal places. (By default, Mathcad shows three decimal places.) The computed solution definitely satisfies the equation.

Keep in mind the following comments on using iterative solve blocks in Mathcad:

- Finding "a" solution does not imply that you have found "the" solution or all solutions. You should always try different initial guesses to check for other solutions. For the sample equation, Mathcad can also solve for the three solutions symbolically. (Try it.) The results are pretty ugly, but the other two roots are imaginary: $-0.74 \pm 3.694i$.
- Solve blocks do support units, but if you are solving for more than one variable, each iterated variable must have the same units. If the variables in your problem do not all have the same units (flow rate and friction loss, for example), you must solve the set of equations without any units on any variable.
- You must provide an initial guess for each iterated variable. (Use an imaginary initial guess to have Mathcad search for imaginary solutions.)

The root( ) Function: An Alternative to Using Solve Blocks

Mathcad provides a `root()` function that can be used to find a single solution to a single equation. The `root()` function does not require the use of a given/find solve block. The `root()` function is an iterative solver, so an initial guess is still required. The procedure is as follows:

define the function	$f(x) := x^3 + 12 \cdot x - 21$
provide a guess	$x := 0$
find a solution	$\text{Soln} := \text{root}(f(x), x)$
display the solution	$\text{Soln} = 1.48$

The `root( )` function will find only one solution. For functions that have multiple solutions, you can provide other guesses (starting values) to search for other roots. To search for imaginary root, use an imaginary guess value.

Finding All Roots of Polynomials

For polynomial functions, Mathcad provides a function that will find all of the roots at one time: the `polyroots()` function. To use the `polyroots()` function, the polynomial coefficients must be written as a column vector, starting with the constant. The polynomial we have been using as an example is

$$x^3 + 0x^2 + 12x - 21 = 0.$$

The $0\,x^2$ term was included in the polynomial as a reminder that the zero must be included in the coefficient vector v.

The coefficients of this polynomial would be written as the column vector

$$v := \begin{pmatrix} -21 \\ 12 \\ 0 \\ 1 \end{pmatrix}$$

The `polyroots()` function could then be used to find all solutions of this polynomial:

$$\text{Soln} := \text{polyroots}(v)$$

$$\text{Soln} = \begin{pmatrix} -0.74 - 3.694i \\ -0.74 + 3.694i \\ 1.48 \end{pmatrix}$$

The `polyroots()` function is a quick way to find all solutions, but only of polynomials.

PROFESSIONAL SUCCESS

Use a QuickPlot to find good initial guesses.

1. Create an X–Y Plot (from the Graphics Toolbar or by pressing [Shift-2])
2. Enter your function on the y-axis.
3. Add a second curve to the plot (select your entire function and then press [comma]), and enter a zero in the y-axis placeholder for the second plot. This will draw a horizontal line across the graph at $y = 0$.
4. Enter the variable used in your function in the placeholder on the x-axis.
5. Adjust the limits on the x-axis as needed to see where your function crosses the $y = 0$ line.

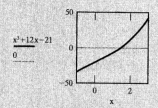
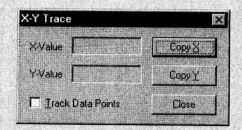

The locations where your function crosses the $y = 0$ line are the roots, or solutions, of the function. This function clearly has a root between 1 and 2, but Mathcad provides a way to get a more accurate value off the graph:

6. Click on the graph to select it.
7. Bring up the X–Y Trace dialog from the Format menu: Format/Graph/Trace. . . .
8. Position the X–Y Trace dialog so that the entire graph is visible.
9. Click on the spot where your function and the $y = 0$ curves cross.
10. Read the x-value at that location from the X–Y Trace dialog box.

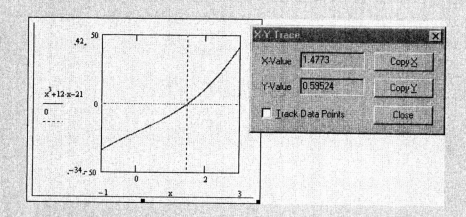

A guessed value of 1.4773 should be very close to the actual root, allowing the iterative solver to converge quickly.

PRACTICE!

Use Mathcad's iterative solver to find solutions for each of the expressions that follow. For each expression, how many solutions should there be? Use different initial guesses to search for multiple solutions.

a. $(x - 3) \cdot (x - 4) = 0$
b. $x^2 - 1 = 0$
c. $x^3 - 2x^2 + 4x = 3$
d. $e^{3x} - 4 = 0$
e. $\sqrt{x^3} + 7x = 10$

APPLICATION: FRICTION LOSSES AND PRESSURE DROP IN PIPE FLOWS

Here's a typical pump-sizing problem:

What size pump (HP) is required to move water at an average velocity of 3.0 ft/s through a 5,000-foot-long pipe with a 1-inch inside diameter? Assume that the viscosity of water is 0.01 poise at room temperature and the pump has an efficiency of 0.70.

Anytime you are designing a piping system, you will need to estimate the friction in the system, since friction can be responsible for much of the pressure drop from one end of the pipe to another. Because of this friction, you need a pump to move the fluid, and you must calculate how big the pump must be—so you have to estimate the friction losses ... to estimate the pressure drop ... to calculate the size of the pump.

There are many contributing factors to pipe friction: valves, bends in the pipe, rough pipe, a buildup of deposits in the pipes, etc. We will consider only the simplest situation: a clean, horizontal, smooth pipe with no valves or bends. For most flows in such a pipe, the Fanning friction factor f can be calculated using the von Karman equation,

$$\frac{1}{\sqrt{\frac{f}{2}}} = 2.5 \ln\left(N_{Re}\sqrt{\frac{f}{8}}\right) + 1.75.$$

NOMENCLATURE

D	Inside pipe diameter.
V_{avg}	Average velocity of the fluid in the pipe.
ρ	Density of the fluid.
μ	Viscosity of the fluid.
L	Length of the pipe.
g_c	Gravitational constant: 32.174 ft lb_m lb_f^{-1} s^{-2} in English units or 1 (no units) in SI.
η	Efficiency of the pump (no units).
P_P	Pump's power rating (HP or kW).
$\dot{m}$	Mass flow rate of fluid in the pipe.

The von Karman equation is valid for smooth pipes (e.g., PVC pipe, not steel pipe) and Reynolds numbers greater than 6,000. The Reynolds number is defined as

$$N_{Re} = \frac{D V_{avg} \rho}{\mu}$$

Once you know the friction factor you can calculate the pressure drop in a horizontal pipe from

$$\Delta P = 4f \frac{L}{D} \frac{\rho (V_{avg})^2}{2 g_c}$$

Once you have the pressure drop, you can determine the energy per unit mass required to overcome friction:

$$h_f = \frac{\Delta P}{\rho}.$$

Then you can determine the pump power required:

$$\eta P_P = h_f \dot{m}$$

With Mathcad, the problem is solved like this:

Information from the Problem Statement:

$V_{avg} := 3 \cdot \frac{ft}{sec}$ $L := 5000 \cdot ft$ $\rho := 1 \cdot in$

$\mu := 0.01 \cdot poise$ $\eta := 0.70$

Commonly Available Data:

$\rho := 1000 \cdot \frac{kg}{m^3}$ Water density

Definition of g_c:

$g_c := 1$ Define in SI, let Mathcad handle units

Calculate the Reynolds Number:

$N_{Re} := \frac{D \cdot V_{avg} \cdot \rho}{\mu}$ Because $N_{Re} > 6,000$, von Karman can be used.

$N_{Re} = 23226$

Solve for the Friction Factor:

$f := 0.001$ Guessed starting value for the iterative solver given

$$\frac{1}{\sqrt{\frac{f}{2}}} = 2.5 \cdot \ln\left(N_{Re} \cdot \sqrt{\frac{f}{8}}\right) + 1.75$$

$f := \text{find}(f)$
$f = 0.0062$ Calculated friction factor

Solve for Pressure Drop:

$$\Delta P := 4 \cdot f \cdot \frac{L}{D} \cdot \frac{\rho \cdot (V_{avg})^2}{2 \cdot g_c} \quad \Delta P = 6.158 \circ \text{atm}$$

$$\Delta P = 90.502 \circ \text{psi}$$

Solve for Energy per Unit Mass Required to Overcome Friction:

$$h_f := \frac{\Delta P}{\rho} \qquad h_f = 623.986 \cdot \frac{N \cdot m}{kg}$$

$$h_f = 208.756 \cdot \frac{ft \cdot lbf}{lb}$$

Solve for Mass Flow Rate:

$$A_{flow} := \pi \cdot \left(\frac{D}{2}\right)^2$$

$$m_{dot} := V_{avg} \cdot A_{flow} \cdot \rho \qquad m_{dot} = 1.668 \cdot 10^3 \cdot \frac{kg}{hr}$$

$$m_{dot} = 3.677 \cdot 10^3 \cdot \frac{lb}{hr}$$

Solve for Required Pump Power:

$$P_p := \frac{h_f \cdot m_{dot}}{\eta} \qquad P_p = 0.413 \circ kW$$

$$P_p = 0.554 \circ hp$$

Surprised by the result? It doesn't take much of a pump just to overcome friction in a well-designed pipeline. (It takes a lot more energy to lift the water up a hill, but that wasn't considered here.) If the pipeline is not designed correctly, the friction losses can change dramatically. That's the subject of Problem 8.7.

8.2 NUMERICAL INTEGRATION

Numerical integration and differentiation of functions are very straightforward in Mathcad. But many times the relationship between the dependent and independent variables is known only through a set of data points. For example, in Chapter 5 we used a data set representing a relationship between temperature and time. (That data set will be presented again shortly.) If you want to integrate the temperature data over time, you have two choices:

- Fit the data with an equation, and then integrate the equation.
- Use a numerical integration method on the data set itself.

Both approaches are common, and both will be described in this section.

8.2.1 Integration

Integrating Functions Numerically

If you have a function, such as the polynomial relating temperature and time that was obtained in Chapter 5, then integrating temperature over time from 0 to 9 minutes is easily performed using Mathcad's definite integral operator. Using the data arrays and the regression statements from before, we have

$$\text{Time} = \begin{bmatrix} & 0 \\ 0 & 0 \\ 1 & 1 \\ 2 & 2 \\ 3 & 3 \\ 4 & 4 \\ 5 & 5 \\ 6 & 6 \\ 7 & 7 \\ 8 & 8 \\ 9 & 9 \end{bmatrix} \qquad \text{Temp} = \begin{bmatrix} & 0 \\ 0 & 298 \\ 1 & 299 \\ 2 & 301 \\ 3 & 304 \\ 4 & 306 \\ 5 & 309 \\ 6 & 312 \\ 6 & 316 \\ 8 & 319 \\ 9 & 322 \end{bmatrix}$$

$$F(x) := \begin{bmatrix} 1 \\ x \\ x^2 \end{bmatrix}$$

b := linfit (Time, Temp, F)

$$b = \begin{bmatrix} 297.721 \\ 1.742 \\ 0.113 \end{bmatrix}$$

$$\int_0^9 (b_0 + b_1 \cdot t + b_2 \cdot t^2)\,dt = 2.777 \cdot 10^3$$

The symbol t was used instead of Time in the integration. The choice is irrelevant, because the integration variable is a dummy variable. Units were not used here. The integration operator does allow units on the limit values, but since linfit() does not support units, the problem is more easily solved without them.

PRACTICE!

Evaluate the following integrals, and check your results using the computational formulas shown to the right of each integral:

- Area under the curve y = 1.5x from x = 1 to x = 3 (a trapezoidal region):

$$\int_1^3 1.5x\,dx \qquad A_{\text{trap}} = \frac{1}{2}(y_{\text{left}} + y_{\text{right}}) \cdot (x_{\text{right}} - x_{\text{left}}).$$

- Volume of a sphere of radius 2 cm:

$$\int_0^{2\text{ cm}} 4\pi r^2\,dr \qquad V_{\text{sphere}} = \frac{4}{3}\pi R^3.$$

- Volume of a spherical shell with inside radius R_i = 1 cm and outside radius R_o = 2 cm

$$\int_{1\text{ cm}}^{2\text{ cm}} 4\pi r^2\,dr \qquad V_{\text{shell}} = \frac{4}{3}\pi(R_o^3 - R_i^3).$$

APPLICATIONS: DETERMINING THE VOLUME OF LIQUID IN A CYLINDRICAL TANK

It is common to need to know how much product you have stored in a partially filled, horizontal, cylindrical tank. The calculation for finding the product, however, is not trivial. We can use Mathcad's ability to integrate functions to solve this problem.

SOME FUNDAMENTALS

- The integral $\int_a^b f(x)dx$ represents the area between the curve f(x) and the x-axis.
- If the f curve is above the x-axis, the calculated area will have a positive sign. A negative sign on the area implies that the curve lies below the x-axis.
- A circle of radius r and centered at the origin can be described by the function $x^2 + y^2 = r^2$.
- A horizontal line (representing the level of liquid in the tank) is described by the function y = constant. We will relate the constant to the depth of liquid in the tank later in this example.

CASE 1: THE TANK IS LESS THAN HALF FULL

When the tank is less than half full, we can compute the volume by multiplying the cross-sectional area of the fluid (shown shaded in the circle at the left in the figure that follows) by the length of the tank, L. The trick is determining the cross-sectional area of the fluid.

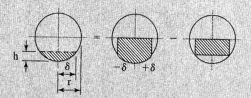

The shaded area in the center circle represents the area between the circle and the x-axis. Note that when the tank is less than half full, there is fluid between $-\delta$ and $+\delta$. We can integrate the formula for the circle between $-\delta$ and $+\delta$ to compute the shaded area in the center circle. Call this area A_1.

The shaded area in the rightmost circle represents the area between the function describing the level of the liquid and the x-axis. The integration limits are again $-\delta$ and $+\delta$. Call this area A_2. Subtracting A_2 from A_1 gives the desired cross-sectional area of the fluid in the tank. All that remains is to carry out these integrations, but first we need to know how δ depends on the level in the tank, h, and the tank radius r.

There is a right triangle involving the level of the liquid and the origin of the axis, shown at the right of the axis. Using the Pythagorean theorem, we can relate the lengths of the sides of the triangle:

$$(h - r)^2 + \delta^2 = r^2$$

Solving for δ yields the first required Mathcad function:

$$\delta(h, r) := \sqrt{r^2 - (r - h)^2}$$

The integral for the area A_1 is then written as a function of r and h as well:

$$A_1(h, r) = -\left[\int_{-\delta(h, r)}^{\delta(h, r)} (-\sqrt{r^2 - x^2})dx\right]$$

Here, the function representing the circle was solved for y, and a negative sign was introduced to calculate the y values below the x-axis. (Since Mathcad's square root operator always returns the positive root, we need to change the sign to obtain the negative y values.) Hence, we have

$$y = -\sqrt{r^2 - x^2}$$

Also, the computed area is below the x-axis, so it will have a negative sign. A minus sign has been included in the function to cause it to return a positive area value.

The integral for A_2 is obtained by integrating the function representing the level of liquid in the tank. If the liquid depth is h, then the level is at a (negative) y value of h−r:

$$A_2(h, r) := -\left[\int_{-\delta(h, r)}^{\delta(h, r)} (h - r)\, dx\right]$$

Again, a minus sign was added to cause the function to return a positive area.

The cross-sectional area of fluid in the tank is then

$$A_{fluid}(h, r) := -\left[\int_{-\delta(h, r)}^{\delta(h, r)} (-\sqrt{r^2 - x^2}) dx - \int_{-\delta(h, r)}^{\delta(h, r)} (h - r) dx \right]$$

or, simplifying slightly,

$$A_{fluid}(h, r) := -\int_{-\delta(h, r)}^{\delta(h, r)} [(-\sqrt{r^2 - x^2}) - (h - r)] dx$$

And the volume in the tank is simply the area times the length:

$$V_{fluid}(h, r, L) := L \cdot \left[-\left[\int_{-\delta(h, r)}^{\delta(h, r)} [(-\sqrt{r^2 - x^2}) - (h - r)] dx \right] \right]$$

CASE 2: THE TANK IS MORE THAN HALF FULL

When the tank is more than half full the procedure is similar

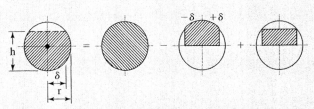

$$A_{fluid} = A_{total} - \int_{-\delta}^{\delta} f_{circle}(x) dx + \int_{-\delta}^{\delta} f_{level}(x) dx$$

The function relating δ to r and h is unchanged, and since the areas are above the x-axis, we don't have to worry about changing the signs. The result is the following function for the volume of the liquid in the tank:

$$V_{fluid}(h, r, L) = L \cdot \left[\pi \cdot r^2 - \int_{-\delta(h, r)}^{\delta(h, r)} [(-\sqrt{r^2 - x^2}) - (h - r)] dx \right]$$

CREATING A GENERAL FUNCTION FOR EITHER CASE

Mathcad's `if()` function can be used to automatically select the appropriate formula:

$$V_{tank}(h, r, L) := L \cdot if\left[(h < r), \left[-\int_{-\delta(h, r)}^{\delta(h, r)} [(-\sqrt{r^2 - x^2}) - (h - r)] dx \right], \right.$$
$$\left. \left[\pi \cdot r^2 - \int_{-\delta(h, r)}^{\delta(h, r)} [(\sqrt{r^2 - x^2}) - (h - r)] dx \right] \right]$$

Here, if $h < r$, then the first formula is used; otherwise the second formula is used. Since both functions work for $h = r$ (a half-full tank), deciding which formula to use at that point is arbitrary.

Integrating Data Sets

Integration via Curve Fitting One approach to integrating a data set is to first fit an equation to the data and then to integrate the equation. This was demonstrated in the previous example: A polynomial was fit to the temperature–time data, and then the polynomial was integrated using Mathcad's definite integral operator.

Integrating without Curve Fitting An alternative to curve fitting is simply to use the data values themselves to compute the integral. This is fairly straightforward if you recall that the integral represents the area between the curve and the x-axis when the data points are plotted.

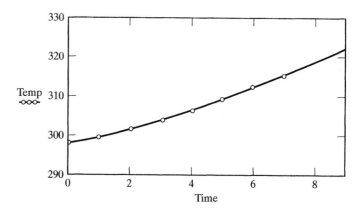

Any method that computes the area under the curve[1] computes the value of the integral. A number of methods are commonly used. One of the simplest divides the area under the curve into a series of trapezoids. The area of each trapezoid is calculated from the data values, and the sum of the areas represents the result of the integration. Since there are 10 data points, there will be nine trapezoids to cover the entire time range. We will keep track of these nine regions by defining a range variable

```
i := 0 .. 8
```

Or, alternatively, we could define the range variable in more general terms as

```
i := 0 .. (last(Time)- 1)
```

We then need a function that computes the area of the leftmost trapezoid:

$$A_0 := \frac{1}{2} \cdot (\text{Temp}_0 + \text{Temp}_1) \cdot (\text{Time}_1 - \text{Time}_0) \qquad A_0 = 298.7$$

We can now generalize this equation to obtain a function capable of calculating the area of any of the nine trapezoids:

$$A_i := \frac{1}{2} \cdot (\text{Temp}_i + \text{Temp}_{i+1}) \cdot (\text{Time}_{i+1} - \text{Time}_i)$$

$$A = \begin{bmatrix} 298.7 \\ 300.424 \\ 302.643 \\ 305.164 \\ 307.916 \\ 310.862 \\ 313.974 \\ 317.235 \\ 320.629 \end{bmatrix}$$

---

[1] Be sure to calculate the area all the way to the x axis ($y = 0$) not just to the $y = 290°$ shown in this figure.

Finally, we sum the area of all of these trapezoids to estimate the total area under the curve. This summation is carried out by using Mathcad's *range variable summation operator* from the Calculus Toolbar:

$$\sum_i A_i = 2.778 \cdot 10^3$$

The result compares well with that computed by using numerical integration of the polynomial in the preceding example. Again, this calculation was performed without using Mathcad's units capability, but units could have been used.

8.2.2 A Trapezoidal Rule Function

We can push this process one step further and write a function that will perform *trapezoidal-rule integration* on any data set:

$$\text{trap}(x, y) := \sum_{i=0}^{\text{length}(x)-2} \frac{y_i + y_{i+1}}{2} \cdot (x_{i+1} - x_i)$$

$$A_{\text{total}} := \text{trap}(\text{Time}, \text{Temp})$$
$$A_{\text{total}} = 2.778 \cdot 10^3$$

In the `trap()` function, the range variable is replaced by defined limits on the summation (Mathcad's standard *summation operator* is used), and the `length()` function is utilized to determine the size of the array, from which the number of trapezoids can be computed. The expressions

```
length(x)-2        used in the trap() function and
last(x)-1          used in the previous example
```

are equivalent as long as the array origin is set at zero, which is assumed in the `trap()` function, since the summation index, i, starts at zero.

PRACTICE!

Create a test data set, and then use the `trap()` function to integrate $y = \cos(x)$ from $x = 0$ to $x = \pi/2$. Vary the number of points in the data set to see how the size of the trapezoids (over the same x range) affects the accuracy of the result. Then check your result using Mathcad's symbolic integrator. The following are the details of the procedure.

- Create the test data set:

```
N_pts := 20
i := 0 .. (N_pts-1)
```

$$x_i := \frac{\pi}{2} \cdot \frac{i}{N_{\text{pts}}-1}$$

$$y_i := \cos(x_i)$$

- Integrate by using the `trap()` function, and vary the number of points in the data set.
- Use symbolic integration to evaluate $\int_0^{\pi/2} \cos(x)\,dx$.

APPLICATIONS: CONTROLLED RELEASE OF DRUGS

When a patient takes a pill, there is a rapid rise in the concentration of the drug in the patient's bloodstream, which then decreases with time as the drug is removed from the bloodstream, often by the kidneys or the liver. Then the patient takes another pill. The result is a time-varying concentration of drug in the blood.

In some situations, there may be therapeutic benefits to maintaining a more constant (perhaps lower) drug concentration for prolonged periods of time. For example, a chemotherapy drug might be active only at concentrations greater than 2 mg/L. The pills for this drug might be designed to raise the concentration in the blood to 15 mg/L to try to keep the concentration above 2 mg/L for as long as possible. If you could keep the concentration of a cancer-fighting drug from falling below 2 mg/L for a month or more, it might do a better job of killing the cancer cells. If you could also reduce the maximum concentration from 15 mg/L to perhaps 10 mg/L, the side effects of the drug might be reduced.

In this example, we will consider an implanted "drug reservoir" for chemotherapy. This drug reservoir is little more than a plastic bag containing a solution of the drug. The shape, materials of construction, and volume of the bag, as well as the concentration of the dissolved drug, can all be varied to change the drug release characteristics. This example considers only the volume and drug concentration in the reservoir.

For preliminary testing of the release characteristics, human subjects would not be used. Instead, the computer model used to generate the accompanying graph assumes that the drug is being released into a body simulator (a 50-liter tank) with a slow (1 mL/min) feed of fresh water and removal of drug solution at the same rate. Three tests were simulated, with the drug reservoir volume and concentration adjusted to give a maximum drug concentration of 10 mg/L in the simulator. The results are shown in the graph (concentration in mg/L, time in hours).

The $conc_1$ curve was produced using a small reservoir containing a high drug concentration. $Conc_3$ used a larger reservoir with a much lower drug concentration. The curves show that, by varying the reservoir volume and concentration, the active period (concentration above 2 mg/liter) of the drug can be adjusted from approximately 800 hours (about 1 month) to almost 1,400 hours (nearly 2 months), without ever causing blood concentrations to exceed 10 mg/L.

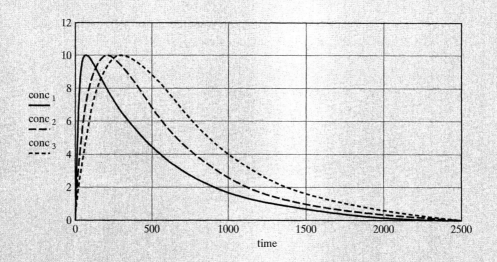

While this simple drug delivery system is a long way from delivering a good, constant concentration of the chemotherapy drug, it does demonstrate that it is quite possible to change the way drugs are administered. By designing better drug delivery systems, we may be able to improve the performance of some drugs and the quality of life of patients.

TOTAL DRUG RELEASE

With the concentration vs. time data, and knowing the flow rate of fluid through the simulator, we can determine the total amount of drug that is released from the reservoir. Multiplying the time by the volumetric flow rate gives the volume that has passed through the simulator. The concentrations can then be plotted against volume (liters). This is shown in the graph that follows, where

```
Q := 60 ml/hr
           time·Q
vol :=    ─────── liters
           1000
```

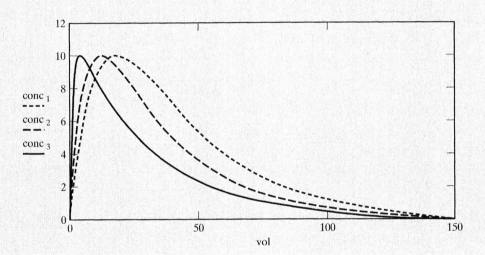

The area under each curve represents the amount of drug released from the associated reservoir in 2,500 hours. The `trap()` function can be used to perform the integrations:

```
D₁ := trap (vol, conc₁)      D₁ = 343 mg
D₂ := trap (vol, conc₂)      D₂ = 446 mg
D₃ := trap (vol, conc₃)      D₃ = 546 mg
```

8.2.3 Simpson's-Rule Integration

Simpson's rule is a popular numerical integration technique that takes three data points, fits a curve through the points, and computes the area of the region below the curve. This operation is repeated for each set of three points in the data set. The common formula for Simpson's rule looks something like

$$A_{total} = \frac{h}{3} \sum_{\substack{all \\ regions}} (y_{i-1} + 4y_i + y_{i+1})$$

where the unusual summation over "all regions" is necessary because an integration region using Simpson's method requires three data points. So, using Mathcad's default array indexing, we see that points 0, 1, and 2 make up the first integration region, points 2, 3, and 4 make up the second region, and so on. The number of integration regions is approximately half the number of data points. The distance between two adjacent points is h. The use of Simpson's rule comes with two restrictions:

- You must have an odd number of data points.
- The independent values (usually called x) must be uniformly spaced, with $h = \Delta x$.

We will look at a way to get around these restrictions later, but first we try an example of applying Simpson's rule when the conditions are met. We create a data set containing seven values and strong curvature:

```
i := 0 .. 6
xᵢ := 1+i
yᵢ := 1+cos(xᵢ)
```

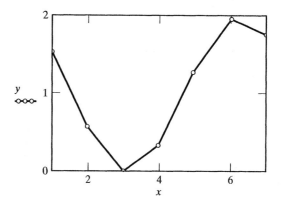

Then we calculate h (from any two x values, since h must be constant) and create a range variable j that will keep track of the index of the point at the center of each integration region:

INTEGRATION REGION, POINT NUMBERS	CENTRAL POINT
0, 1, 2	1
2, 3, 4	3
4, 5, 6	5

```
h := x₁-x₀        h = 1
j := 1, 3 .. 5    j =
                  ┌─┐
                  │1│
                  │3│
                  │5│
                  └─┘
```

Next, we apply Simpson's rule to determine the area under the curve, again by using the range variable summation operator:

$$A_{Simpson} := \frac{h}{3} \cdot \sum_j (y_{j-1} + 4 \cdot y_j + y_{y+1})$$

$$A_{Simpson} = 5.814$$

Now we can compare the results with trapezoidal-rule integration and exact integration of the cosine function:

$$A_{trap} := \text{trap}(x, y) \qquad A_{trap} = 5.831$$

$$A_{exact} := \int_1^7 (1+\cos(x))dx \qquad A_{exact} = 5.816$$

Simpson's method came a lot closer to the exact result than trapezoidal integration did—it usually does. Because Simpson's rule connects data points with smooth curves rather than straight lines, it typically fits data better than the trapezoidal rule does. But Simpson's rule has those two restrictions that make it useless in many situations. Is there a way to get around these restrictions? Yes, there is: You can use a cubic spline to fit any data set with a smooth curve and then use cubic spline interpolation to compute a set of values that covers the same range as the original data, but has an odd number of data points and uniform point spacing. After that, you can use Simpson's rule on the interpolated data.

To test this approach, we will use the temperature–time data employed in the previous examples. These data consist of 10 uniformly spaced values. The uniform spacing is good, but 10 values won't work with Simpson's rule.[2] A spline fit to the data was performed in Chapter 5:

Time =

	0
0	0
1	1
2	2
3	3
4	4
5	5
6	6
7	7
8	8
9	9

Temp =

	0
0	298
1	299
2	301
3	304
4	306
5	309
6	312
7	316
8	319
9	322

```
vs := cspline(Time, Temp)
```

Then we can use the `interp()` function to find temperature values at 11 points over the same time interval, 0 to 9 minutes:

```
i := 0 .. 10     eleven values
```

---

[2]Another common way to get around the odd-number-of-data-points restriction is to use Simpson's rule as far as possible and then, if there is an even number of points, finish the integration by using a trapezoid for the last two points.

$$t_i := \frac{9 \cdot i}{10}$$

$$t = \begin{array}{|c|c|} \hline & 0 \\ \hline 0 & 0 \\ \hline 1 & 0.9 \\ \hline 2 & 1.8 \\ \hline 3 & 2.7 \\ \hline 4 & 3.6 \\ \hline 5 & 4.5 \\ \hline 6 & 5.4 \\ \hline 7 & 6.3 \\ \hline 8 & 7.2 \\ \hline 9 & 8.1 \\ \hline 10 & 9 \\ \hline \end{array}$$

```
Temp_interp_i := interp(vs, Time, Temp, t_i)
```

$$\text{Temp}_{\text{interp}} = \begin{array}{|c|c|} \hline & 0 \\ \hline 0 & 298 \\ \hline 1 & 299.225 \\ \hline 2 & 301.003 \\ \hline 3 & 303.093 \\ \hline 4 & 305.401 \\ \hline 5 & 307.893 \\ \hline 6 & 310.538 \\ \hline 7 & 313.32 \\ \hline 8 & 316.25 \\ \hline 9 & 319.24 \\ \hline 10 & 322.358 \\ \hline \end{array}$$

Now we use Simpson's rule on `t` and `Temp_interp` instead of `Time` and `Temp`:

```
j := 1, 3 .. 9
h := t_1 - t_0
```

$$A_{\text{Simpson}} := \frac{h}{3} \cdot \sum_j \left(\text{Temp}_{\text{interp}_{j-1}} + 4 \cdot \text{Temp}_{\text{interp}_j} + \text{Temp}_{\text{interp}_{j+1}} \right)$$

$$A_{\text{Simpson}} = 2.777 \cdot 10^3$$

The result compares well with that obtained from the trapezoidal integration, which is not surprising for this data set, since it does not show a lot of curvature. (There is not a lot of difference between connecting the points with lines or curves for the data set, so there is little difference between the results computed by the two methods.)

8.3 NUMERICAL DIFFERENTIATION

Evaluating Derivatives of Functions Numerically

Mathcad can take the derivative of a function such as

$$y = b_0 e^{b_1 t}$$

If you use the live symbolic operator (→) to evaluate an expression symbolically, Mathcad will use its symbolic processor and give you another function:

$$\frac{d}{dt}(b_0 \cdot e^{b_1 \cdot t}) \rightarrow b_0 \cdot b_1 \cdot \exp(b_1 \cdot t)$$

Or if the values of b_0, b_1, and t are specified before evaluating the derivative, Mathcad's symbolic processor will calculate a numeric result:

$$b_0 := 3.4$$
$$b_1 := 0.12$$
$$t := 16$$
$$\frac{d}{dt}(b_0 \cdot e^{b_1 \cdot t}) \rightarrow 2.7829510554706259326$$

The extreme number of significant figures is a reminder that Mathcad solved for the value using the symbolic processor. The default number of digits displayed is 20.

On the other hand, if you use the "numerical evaluation" symbol (a plain equal sign, =), then Mathcad will use its numeric processor to calculate the value of the derivative:

$$b_0 := 3.4$$
$$b_1 := 0.12$$
$$t := 16$$
$$\frac{d}{dt}(b_0 \cdot e^{b_1 \cdot t}) = 2.783$$

If you use the numeric processor, you must specify the values of b_0, b_1, and t before Mathcad evaluates the derivative.

Derivatives from Experimental Data

If the relationship between your variables is represented by a set of values (a data set), rather than a mathematical expression, you have two choices for trying to determine the derivative at some specified point:

- Fit the data with a mathematical expression and then differentiate the expression at the specified value.
- Use numerical approximations for derivatives on the data points themselves. (This approach is only practical when you have "clean" data.)

8.3.1 Using a Fitting Function

Curve fitting was covered in Chapter 5, so the process is only summarized here:

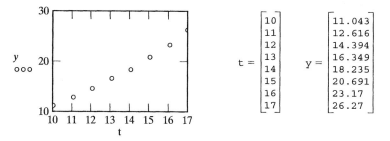

$$t = \begin{bmatrix} 10 \\ 11 \\ 12 \\ 13 \\ 14 \\ 15 \\ 16 \\ 17 \end{bmatrix} \quad y = \begin{bmatrix} 11.043 \\ 12.616 \\ 14.394 \\ 16.349 \\ 18.235 \\ 20.691 \\ 23.17 \\ 26.27 \end{bmatrix}$$

The data are expected to fit the exponential model

$$y = b_0 e^{b_1 t},$$

which can be rewritten in linear form as

$$\ln(y) = \ln(b_0) + b_1 t.$$

We can then use linear regression and a bit of math to determine b_0 and b_1:

```
in := intercept(t, ln(y))            in = 1.187
sl := slope(t, ln(y))                sl = 0.123
b₀ := e^in                           b₀ = 3.276
b₁ := sl                             b₁ = 0.123
```

Once we have a mathematical expression, we can evaluate the derivative at $t = 16$ by using either the symbolic or numeric processor, as described earlier. The results of using the numeric processor are as follows:

```
b₀ := e^in                           b₀ = 3.276
b₁ := sl                             b₁ = 0.123
t := 16
d/dt (b₀ · e^(b₁t)) = 2.861
```

8.3.2 Using Numerical Approximations for Derivatives

The derivative physically represents the slope of a plot of the data at a specified point. You can approximate the derivative at any point by estimating the slope of the graph at that point. Just as there are several ways to estimate the slope, there are several ways to compute numerical approximations for derivatives. Since array index values are used in these calculations, here are the data (again), along with the array index values:

$$i = \begin{bmatrix} 0 \\ 1 \\ 2 \\ 3 \\ 4 \\ 5 \\ 6 \\ 7 \end{bmatrix} \quad t = \begin{bmatrix} 10 \\ 11 \\ 12 \\ 13 \\ 14 \\ 15 \\ 16 \\ 17 \end{bmatrix} \quad y = \begin{bmatrix} 11.043 \\ 12.616 \\ 14.394 \\ 16.349 \\ 18.235 \\ 20.691 \\ 23.17 \\ 26.27 \end{bmatrix}$$

One way to estimate the slope at point i = 6 (where t = 16) would be

$$\left.\frac{dy}{dt}\right|_{i=6} \approx \frac{y_7 - y_6}{t_7 - t_6}$$

If that equation is valid, then the following equation is equally valid (both are approximations):

$$\left.\frac{dy}{dt}\right|_{i=6} \approx \frac{y_6 - y_5}{t_6 - t_5}.$$

The first expression uses the point at t = 16 and the point to the right (i = 7 or t = 17) to estimate the slope and is called a *forward finite-difference approximation for the first derivative* at i = 6 The term *finite difference* is used because there is a finite distance between the *t* values used in the calculation. Finite-difference approximations are truly equal to the derivatives only in the limit as Δt approaches zero.

The second expression uses the point at t = 16 and the point to the left (t = 15) to estimate the slope and is called a *backward finite-difference approximation for the first derivative* at i = 6. You can also write a *central finite-difference approximation for the first derivative* at i = 6:

$$\left.\frac{dy}{dt}\right|_{i=6} \approx \frac{y_7 - y_5}{t_7 - t_5}.$$

Central differences tend to give better estimates of the slope and are the most commonly used. Applying the preceding equation to calculate the derivative at point i = 6 (t = 16) in the data set, we find that it has a value of 2.79:

$$\text{slope}_{16} := \frac{y_7 - y_5}{t_7 - t_5}$$

$$\text{slope}_{16} = 2.79$$

There are also finite-difference approximations for higher order derivatives. For example, a *central finite-difference approximation for a second derivative* at i = 6 can be written (assuming uniform point spacing—i.e., Δt constant) as

$$\left.\frac{d^2y}{dt^2}\right|_{i=6} \approx \frac{y_7 - 2y_6 + y_5}{(\Delta t)^2}$$

PRACTICE!

Use central and forward difference approximations to estimate dy/dx at x = 1. Try both the clean and noisy data sets.

$$x := \begin{bmatrix} 0.0 \\ 0.5 \\ 1.0 \\ 1.5 \\ 2.0 \\ 2.5 \\ 3.0 \end{bmatrix} \quad y_{\text{clean}} := \begin{bmatrix} 0.00 \\ 0.48 \\ 0.84 \\ 1.00 \\ 0.91 \\ 0.60 \\ 0.14 \end{bmatrix} \quad y_{\text{noisy}} := \begin{bmatrix} 0.24 \\ 0.60 \\ 0.69 \\ 1.17 \\ 0.91 \\ 0.36 \\ 0.18 \end{bmatrix}$$

A polynomial can be fit to the noisy data as follows:

$$f(x) := \begin{bmatrix} x \\ x^2 \\ x^3 \end{bmatrix} \quad b := \text{linfit}(x, y_{\text{noisy}}, f) \quad b = \begin{bmatrix} 1.434 \\ -0.588 \\ 0.041 \end{bmatrix}$$

Try using the differentiation operator on the Calculus Toolbox to evaluate

$$\frac{d}{dx}(1.434x - 0.588x^2 + 0.041x^3)$$

at $x = 1$.

Note: The derivative values you calculate in this "Practice!" box will vary widely. Calculating derivatives from noisy data is highly prone to errors, and you should try to avoid doing that if possible. If you must take derivatives from experimental data, try to get good, clean data sets, or use regression to find the best fit curve through the data and find the derivative from the regression result.

SUMMARY

In this chapter, you learned to use several standard numerical methods in Mathcad, including an iterative solve block and methods for numerical integration and differentiation. The numerical integration techniques involved using Mathcad's integration operators (from the Calculus Toolbox) when you were working with a function, and numerical techniques like trapezoidal- or Simpson's-rule integration when you were working with a data set. Similarly, Mathcad's differentiation operators apply when you need to take the derivative of a function, and finite difference methods were presented for evaluating derivatives when you have a data set.

MATHCAD SUMMARY

Iterative Solutions:

given
: The keyword that begins an iterative solve block. Remember to specify an initial guessed value before using this keyword.

find()
: The keyword that closes an iterative solve block and the function that performs the iteration and returns the solution. Remember to assign the value returned by find() to a variable.

root(f(x, y), x)
: The root() function uses an iterative solver to find a single root of a function of one or more variables. Before using the root() function, the function to be solved must be defined and an initial guess for the iteration variable must be set. The root() function takes two arguments: the function to be solved, f(x,y), and the iteration variable x.

polyroots(v)
: The polyroots() function returns all roots of a polynomial. The coefficients of the polynomial are sent to the polyroots() function as a column vector, with the constant in the polynomial as the first element of the vector.

Integration:

of a function	Use Mathcad's integration operators from the Calculus Toolbox.
of a data set	Trapezoidal and Simpson's rules can be used to approximate the integral, or you can fit a function to the data and then integrate the function.

Differentiation:

of a function	Use Mathcad's differentiation operators from the Calculus Toolbar.
of a data set	Finite differences can be used to estimate the values of the derivative, or you can fit a function to the data and then differentiate the function. (The latter approach is preferred if you have "noisy" data.)

KEY TERMS

curve fitting
finite difference
initial value (guess)
iteration
iterative solve block

iterative solver
numerical differentiation
numerical integration
polynomial coefficients

Simpson's Rule
solve block
tolerance
trapezoidal rule

Problems

1. **FINDING SOLUTIONS WITH THE root() FUNCTION**
   Use the root() function to find the solution(s) of the following equations:
   a. $x - 3 = 0$ (admittedly, this one is kind of obvious)
   b. $x^2 - 3 = 0$
   c. $x^2 - 4x + 3 = 0$

2. **FINDING INTERSECTIONS**
   Use the root() function to find the intersection(s) of the listed curves (Part a has been completed as an example):
   a. $y = x^2 - 3$ and $y = x + 4$

   The intersections are at the locations where the y values are equal, so start by eliminating y from the equations:

   $$x^2 - 3 = x + 4.$$

   Next, get all terms on one side of the equation, so that the equation is set equal to zero:

   $$x^2 - x - 7 = 0.$$

   Create a QuickPlot to find good initial guesses for x, then use the root() function to find more precise x values at each intersection:

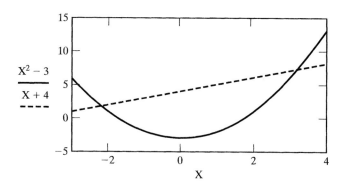

```
guess:    x := -2
          root(x² - x - 7, x)  =  -2.193
          x := 3
          root(x² - x - 7, x)  =  -3.193
```

b. $y = x^2 - 3$ and $y = -(x^2) + 4$

c. $y = x^2 - 3$ and $y = e^{-x}$

3. **FINDING THE ZEROES OF BESSEL FUNCTIONS**

   Bessel functions are commonly used when solving differential equations in cylindrical coordinates. Two Bessel functions, J_0 and J_1, are shown in the following graph[3]

   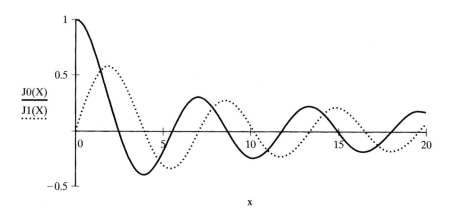

   These Bessel functions are available as built-in functions in Mathcad and are called `J0( )` and `J1( )`. Use the `root( )` function with different initial guesses to find the roots of $J0(x)$ and $J1(x)$ in the range $0 < x < 20$.

   Note: The `J0( )` function has a zero in its name, not the letter "o."

4. **FINDING POLYNOMIAL ROOTS**

   Use the `polyroots()` function to find all roots for the following polynomial expressions:

---

[3]More precisely, these are Bessel functions of the first kind, of order zero and one, respectively.

a. $x - 3 = 0$ (admittedly, not much of a polynomial)
b. $x^2 - 3 = 0$
c. $x^2 - 4x + 3 = 0$
d. $x^3 - 3x^2 + 4x - 12 = 0$

5. **FINDING INTERSECTIONS OF POLYNOMIALS**

   Use the `polyroots()` function to find all intersections of the following polynomials (part a has been completed as an example):

   a. $y = x^2 - 3$ and $y = x + 4$

   Eliminate y from the equations:

   $$x^2 - 3 = x + 4$$

   Then, get all terms on one side of the equation:

   $$x^2 - x - 7 = 0.$$

   Create a column vector containing the coefficients of the polynomial (with the constant on top). Then use the `polyroots()` function to find each intersection:

   $$v := \begin{pmatrix} -7 \\ -1 \\ 1 \end{pmatrix}$$

   $$\text{polyroots}(v) = \begin{pmatrix} -2.193 \\ 3.193 \end{pmatrix}$$

   b. $y = x^2 - 3$ and $y = -(x^2) + 4$

   c. $y = x^2 - 3$ and $y = 2x^3 - 3x^2 + 4x - 12$

   $y = 2x^3 - 3x^2 + 4x - 12$ and $y = -(x^2) + 4$

6. **ITERATIVE SOLUTIONS**

   Use a solve block (given/find) to find the roots of the following equations (you may want to use a QuickPlot to find out how many roots to expect):

   a. $(x - 5) \cdot (x + 7) = 0$

   b. $x^{0.2} = \ln(x)$

   c. $\tan(x) = 2.4x^2$ search for roots in the range $-2 < x < 2$

7. **FRICTION LOSSES AND PRESSURE DROP IN PIPE FLOWS**

   In a Mathcad Application section, the friction loss in a well-designed pipeline was determined. What if the pipeline is not well designed? That's the subject of this problem.

   When you are moving water in a pipeline, the flow velocity is typically around 3 ft/s—the value that was used in the Application problem in Section 8.1. This is a commonly used velocity that allows you to move a reasonable amount

of water quickly and without too much friction loss. What happens to the friction loss and the required pump power when you try to move water through the same pipeline at a velocity of 15 ft/s?

8. **REAL GAS VOLUMES**

The Soave–Redlich–Kwong (SRK) equation of state is a commonly used equation that relates the temperature, pressure, and volume of a gas under conditions when the behavior of the gas cannot be considered ideal (e.g., moderate temperature and high pressure). The equation is

$$P = \frac{RT}{(\hat{V} - b)} - \frac{\alpha a}{\hat{V}(\hat{V} + b)},$$

where α, a, and b are parameters specific to the gas, R is the ideal gas constant, P is the absolute pressure, T is absolute temperature, and

$$\hat{V} = \frac{V}{n}$$

is the molar volume (volume per mole of gas).

For common gases, the parameters α, a, and b can be readily determined from available data. Then, if you know the molar volume of the gas, the SRK equation is easy to solve for either pressure or temperature. However, if you know the temperature and pressure and need to find the molar volume, an iterative solution is required.

Determine the molar volume of ammonia at 300°C and 1,200 kPa by

a. using the ideal gas equation.
b. using the SRK equation and an iterative solve block.

For ammonia at these conditions;

$$\alpha = 0.7007;$$
$$a = 430.9 \text{ kPa} \cdot \text{L}^2/\text{mole}^2;$$
$$b = 0.0259 \text{ L/mole}.$$

9. **REQUIRED SIZE FOR A WATER RETENTION BASIN**

In urban areas, as fields are turned into streets and parking lots, water runoff from sudden storms can become a serious problem. To prevent flooding, retention basins are often built to hold excess water temporarily during storms. In arid regions, these basins are dry most of the time, so designers sometimes try to incorporate alternative uses into their design. A proposed design is a half-pipe for skateboarders that can hold 100,000 cubic feet of water during a storm. The radius of the half-cylinder needs to be scaled to fit the skateboarders, and a radius of 8 feet is proposed.

a. What is the required length of the basin to hold the 100,000 ft³ of storm water? (There may be several short sections of half-pipes to provide the required total volume.)
b. If the basin is filled with water to a depth of 4 feet after a storm, what volume of water is held in the basin?

10. **WORK REQUIRED TO STRETCH A SPRING**

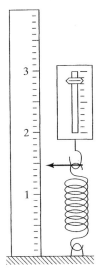

The device shown in the accompanying figure can be used to determine the work required to extend a spring. This device consists of a spring, a spring balance, and a ruler. Before stretching the spring, its length is measured and found to be 1.3 cm. The spring is then stretched 0.4 cm at a time, and the force indicated on the spring balance is recorded. The resulting data set is shown in the following table:

MEASUREMENT (CM)	UNEXTENDED LENGTH (CM)	EXTENDED LENGTH (CM)	FORCE (N)
1.3	1.3	0.0	0.00
1.7	1.3	0.4	0.88
2.1	1.3	0.8	1.76
2.5	1.3	1.2	2.64
2.9	1.3	1.6	3.52
3.3	1.3	2.0	4.40
3.7	1.3	2.4	5.28
4.1	1.3	2.8	6.16
4.5	1.3	3.2	7.04
4.9	1.3	3.6	7.92

Work can be computed as

$$W = \int F\, dx,$$

where x is the extended length of the spring.

a. Calculate the work required to stretch the spring from an extended length of 0 cm to 3.6 cm. Watch the signs on the forces in this problem. Force is a vector quantity, so there is an associated direction.

b. Is the force being used to stretch the spring in the same direction as the movement of the spring or in the opposite direction?

11. WORK REQUIRED TO STRETCH A NONLINEAR SPRING

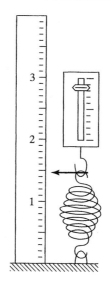

EXTENDED LENGTH (CM)	FORCE (N)
0.0	0.00
0.4	0.15
0.8	0.41
1.2	0.72
1.6	1.07
2.0	1.46
2.4	1.89
2.8	2.35
3.2	2.83
3.6	3.33

Springs come in a variety of sizes and shapes. The shape can sometimes affect the performance of the spring. The spring shown in the preceding figure pulls easily at first, but gets stiffer the more it is stretched. (See the accompanying table.) It is a nonlinear spring. Calculate the work required to stretch this spring

a. from an extended length of 0 cm to 1.6 cm.
b. from an extended length of 2.0 cm to 3.6 cm.

12. WORK REQUIRED TO EXPAND A GAS

In Problems 7.10 and 7.11, work was calculated as

$$W = \int F \, dx.$$

Because force divided by area equals pressure, and length times area equals volume, work can also be found as

$$W = \int P \, dV.$$

This form is handier for dealing with gas systems like that shown in the accompanying figure. As the piston is lifted, the pressure in the sealed chamber will fall. By monitoring the change in position on the ruler and knowing the cross-sectional area of the chamber, we can calculate the chamber volume at each pressure. Care must be taken to allow the system to equilibrate at room temperature before taking the readings. The pressure and volume after equilibrium are listed in the accompanying table.

VOLUME (ML)	PRESSURE (ATM)
1.3	4.50
1.7	3.44
2.1	2.79
2.5	2.34
2.9	2.02
3.3	1.77
3.7	1.58
4.1	1.43
4.5	1.30
4.9	1.19

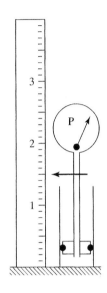

a. Calculate the work required to expand the gas from a volume of 1.3 mL to 4.9 mL.

b. Is this work being done by the gas on the surroundings or by the surroundings on the gas?

13. CALCULATING SPRING CONSTANTS

The extension of a linear spring is described by Hooke's law,

$$F = kx,$$

where x is the extended length of the spring (i.e., the total length of the stretched spring minus the length of the spring before stretching) and k is the spring constant. The spring constant quantifies the "stiffness" of the spring. Hooke's law can also be written in differential form as

$$\frac{dF}{dx} = k.$$

For a linear spring, dF/dx is a constant, the *spring constant*. For a nonlinear spring, dF/dx will not be constant, but taking this derivative at various spring extensions can help you see how the spring characteristics change as the spring is pulled.

Use the data from Problems 8.10 and 8.11 to compute

a. the spring constant for the linear spring (Problem 8.10).

b. the derivative dF/dx as a function of x for the nonlinear spring (Problem 8.11). Use a central difference approximation for the derivative whenever possible, and create a plot of dF/dx vs. x.

14. THERMAL CONDUCTIVITY

    Thermal conductivity is a property related to a material's ability to transfer energy by conduction. Good conductors, like copper and aluminum, have large thermal conductivity values. Insulating materials should have low thermal conductivities to minimize heat transfer.

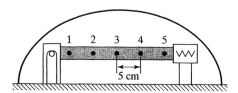

The preceding figure illustrates a device that could be used to measure thermal conductivity. A rod of the material to be tested is placed between a resistance heater (on the right) and a cooling coil (on the left). Five (numbered) thermocouples have been inserted into the rod at 5-cm intervals. The entire apparatus is placed in a bell jar, and the air around the rod is pumped out to reduce heat losses.

To run the experiment, a known amount of power is sent to the heater, and the system is allowed to reach steady state. Once the temperatures are steady, the power level and temperatures are recorded. A data sheet might look like the following:

ROD DIAMETER:	2 CM	
TC SPACING:	5 CM	
POWER:	100 WATTS	
	TC #	TEMP. (K)
	1	348
	2	387
	3	425
	4	464
	5	503

The thermal conductivity can be determined from Fourier's law,

$$\frac{q}{A} = -k\frac{dT}{dx}.$$

where q is the power being applied to the heater and A is the cross-sectional area of the rod.

Note: The term q/A is called the *energy flux* and is a vector quantity—that is, it has a direction as well as a magnitude. As drawn, with the energy source on the right, the energy will be flowing in the −x direction, so the flux in this problem is negative.

a. Use finite-difference approximations to estimate dT/dx at several locations along the rod, and then calculate the thermal conductivity of the material.
b. Thermal conductivity is a function of temperature. Do your dT/dx values indicate that the thermal conductivity of the material changes appreciably between 348 and 503 K?

15. CALCULATING WORK

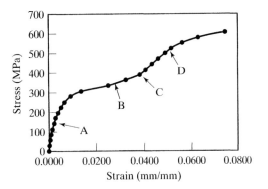

In the previous chapter, there was a note in the "Application: Analyzing Stress–Strain Diagrams" box to the effect that an alternative method of integrating a force–displacement graph is available to calculate work. The alternative method is numerical integration using a function such as trap(). The stress–strain data (shown in the accompanying graph) for a composite material have been converted to force–displacement data, abridged, and tabulated as follows:

F (N)	X (MM)
0	0.000
5500	0.009
8300	0.014
13800	0.023
16500	0.033
22100	0.051
24800	0.066
30300	0.137
33100	0.248
38600	0.380
41400	0.407
46900	0.461
49600	0.488
55200	0.561
57900	0.628
60700	0.749

Calculate the work done on the sample.

Index

A

Abbreviations
    units, 21–24
Absolute value, 53
acosh(z), 49
acos(z), 48
Add Line button, 145
Advanced math functions, 50–51
    discontinuous functions, 52
    if() function, 51, 52, 56
    random-number-generating
        function, 52
Alternative fuel, 141, 156
Anchor point, 11
Antoine's equation, 61
Apparent density, 62
APPEND() function, 53
APPENDPRN() function, 53
Argument list, 143
Arrays, 67–68, 94
    addition/subtraction, 87
    choosing a subset of, 86
    combining, 86
    copying/pasting from other
        Windows programs, 78–79
    deleting rows/columns from, 82
    functions, 42–45, 47–56, 94–96
    initializing, 68–79
    inserting row/column into, 80, 82
    matrix() function, 77
    maximum size of, 68
    modifying, 79–87
    operations, 87–94
    origin, 68
    range variables, 77
    reading values from a text file, 70
    selecting a portion of, 82
    selecting a single row of, 85
    two-dimensional array, 77
    typing values into, 68
    units on matrix elements, 77
asinh(z), 49
asin(z), 48
Assignment operation (:=), 11
atan(z), 48
atanh(z), 49

augment() function, 86
Automatic recalculation, 3–4
Average fluid velocity, 39

B

Backward finite difference
    approximation, 237
Binary, 14
Biomedical engineering, 6, 216
Block iterative solve, 218
Boolean (logical) operators, 52
Brackets
    use of in text, 6
Break statement, 167–168
Bridge
    suspension designing, 41, 64
    Wheatstone, 105, 106

C

C programming language
    problem solving ability, 1–3
Calculus Toolbar, 203, 204, 209, 193
    derivative button, 203
    nth derivative button, 203
    operations using, 207–209
Cautions on unit names, 21
ceil(x), 50
Central finite difference approximation,
    237
Changing a value or a variable name, 16
Changing an operator, 16
Cladding of fibers, 55
Coefficient of determination, R^2, 122
CO_2 emissions
    reducing, 31, 32, 39
Collect, 190
Collect operation, 190, 207
cols(A) function, 94
Column operator, 82, 83
Common math operators, 44
Component, 74
Composite materials, 6, 181, 204
concat(S1,S2), 53
Conditional execution, 148, 161
Continue statement, 168–169

Conventions used in text, 6–7
Copying and pasting arrays, 78–79
corr() function, 122, 132
Correlation coefficient, 122
cos(z), 47
coth(z), 49
cot(z), 47
csort() function, 95
csort(A,n), 95
cspline() function, 119, 120, 132
Cubic spline interpolation, 118–120
Current, 20, 21
Curve fitting, 122–131, 236
    generalized linear regression,
        124–127
    integration via, 226
    integration without, 227
    other linear models, 127, 128
    simple linear regression, 122–12

D

Data, 148
Data analysis, 4
Data analysis functions, 109–133
Decimal, 14
Definite integrals, 197, 198, 199,
    201–203, 208
    mixed limits with multiple
        integration variables, 198
    numerical evaluation, 198
    symbolic evaluation with mixed
        limits, 198
    symbolic evaluation with numer
        limits, 198
Deleting rows or columns, 82
Derivative button
    Calculus Toolbar, 203, 204, 209,
Derivatives, 4, 209
    from experimental data, 233
    using a fitting function, 234
    using numerical approximation
        234–237
Design tool
    Mathcad as, 3–4
Determinant, 91–93, 193
Determinant operation, 193

Differentiation, 203–205
    with respect to multiple variables, 204
Dimensional equation, 61
Discontinuous functions, 52
    Kronecker Delta, 52
    Heaviside Step Function, 52
Display result (=), 12
Displaying a value, 12
Drugs
    controlled release of, 229

E

e (predefined variable), 13–14
Edit cursor, 10, 11
Editing equations, 14–19
    changing an operator, 16
    changing a value or a variable name, 16
    highlighting a region, 17
    inserting a minus sign, 16–17
    selecting part of an equation, 14–15
Editing session
    sample, 29, 30
Editing the units on a value or result, 22
Elementary math functions and operators, 44, 45, 47
    common math operators, 44
    logarithm and exponentiation functions, 47–49
    QuickPlot and visualization of functions, 49
Element-by-element multiplication, 88, 89
Element-by-element plotting, 110–113
Elements, 67
Ellipsis, 76
Energy and work, 21
Engineering in medicine, 216
Engineer's tool kit and Mathcad, 5–6
Equal signs, four types, 11
    Assignment (:=), 12
    Display result (=), 12
    Symbolic equality (=), 12
    Global assignment (≡), 12
Equation edit mode, 13
Equation region, 13
Equations,
    changing an operator, 16–20
    changing a value/variable name, 16
    collecting terms, 186, 190
    editing, See Editing equations
    entering, 14–20
    expanding, 186–187
    exponents, 14
    factoring, 186, 187–188
    inserting into other software programs, 5
    manipulating, 186–191, 193
    nondecimal values, entering, 14–17, 19, 20
    partial-fraction expansion, 190–191
    predefined variables, 13, 14
    selecting part of, 15
    simplifying, 186, 189–190
    solving symbolically, 186, 187
    substituting, 186, 188–189
Error trapping, 164
Evaluation and Boolean Toolbar, 10
Evaluation Style dialog box
    Symbolics menu, 185, 186
Examples
    conventions used in text, 6–7
Expanding variables, 186–187
Expand operation, 187, 188, 190, 207
expfit( ), 129
expfit(vx, vy, vg), 129
Exponential, 129
Exponential curve, 128
Exponents, 15
$exp(x)$, 48
Extractor
    material balances on, 101, 103

F

Factorial, 44
Factoring an expression, 186
Factor operation, 187, 188, 207
Fiber classing, 55
Fiber optics
    Mathcad applied to, 55
File-handling functions, 53
find() keyword, 218
Finite difference
    approximations of derivatives, 233
    defined, 237
Finite-difference approximation, 235
Flash, 103
Flash distillation, 103
floor(x) function, 50
Flowcharts, 145–148
Font Bar, 10
Force, 21
Force components and tension in wires, 63
For loops, 166–167
Format Bar, 10
Format Menu, 10
Format Number dialog, 14
Format Result dialog, 14
Format Text dialog box, 28
Format Toolbar 10
Format/Style.../Normal/Modify, 29
Forward finite difference approximation, 237
Function, 42–45, 47–56, 94–96
    acosh(z), 49
    acos(z), 48
    APPEND(), 53
    APPENDPRN(), 53
    asinh(z), 49
    asin(z), 48
    atan(z), 48
    atanh(z), 49
    augment(), 86
    ceil(x), 50
    cols(A), 94
    concat(S1,S2), 53
    corr(), 122, 132
    cos(z), 47
    coth(z), 49
    cot(z), 47
    csort(), 95
    cspline(), 119, 120, 132
    exp(x), 48
    expfit(), 129
    floor(x), 50
    identity(), 79
    if(), 51, 52, 56
    intercept(), 122, 132
    interp(), 119–121, 132
    last(v), 94, 115
    length(v), 94, 122, 129
    lgsfit(vx, vy, vg), 129
    linfit(), 124, 129
    linterp(), 118–119, 132
    ln(z), 45
    log(z), 45
    logfit(vx, vy, vg), 129
    lspline(), 119–121, 132
    matrix(), 77
    max(A), 94
    min(A), 94
    mean() function, 46
    mod(x,y), 50
    num2str(x), 53
    polyroots(), 219
    pspline(), 119–121, 132
    pwrfit(vx, vy, vg), 129
    READ (), 53
    READPRN (), 53, 70
    reverse(v), 94
    root( ), 219
    rows(A), 94
    rsort(), 94–95

sech(z), 49
sec(z), 47
sin(z), 47
sinfit(vx, vy, vg), 129
sinh(z), 49
slope(), 132
sort(v), 94
stack(), 86
stdev(A), 117
Stdev(A), 118
str2num(s), 53
strlen(s), 53
submatrix(), 86
substr(S,n,m), 53
tan(z), 47
tanh(z), 49
trail(), 51, 53
var(A), 117
Var(A), 118
WRITE(), 53
WRITEPRN(), 53
Functions, 148
    advanced math, 50–51
    arrays, 94–100
    conventions used in text, 6
    data analysis, 109–133
    discontinuous, 52
    elementary math functions and operators, 44–47
    file-handling, 53
    hyperbolic trigonometric, 49
    inverse hyperbolic trigonometric, 49
    inverse trigonometric, 48
    logarithm and exponentiation, 45–47
    parameters, 43
    programming, 169–170
    random-number-generating, 52
    round-off and truncation, 50, 51
    scalar, 43, 44, 53
    standard trigonometric, 48
    statistical, 117, 118
    string, 53
    symbolic math, 181–183, 185–195, 197–199, 201
    trigonometric, 47–50
    user-written, 47–49, 53–54
    validating while developing, 48

G

g (gravitational acceleration), 13–14, 21
Gas absorber
    material balances on, 101, 103
Generalized linear regression, 124–127
Given keyword, 218

Global assignment (≡), 12
Graph Toolbar, 10
Graphing, 110–114
    element against element, 110–113
    modifying graphical display attributes, 115–117
    plotting multiple curves, 112
    vector against vector, 110–112
    QuickPlot, 114, 115
Greek Symbol Toolbar, 10, 13

H

Handles, 29
Heaviside step function, 52
Hexadecimal values, 14
    entering, 14–17, 19, 20
Highlighted region, 16
Hooke's law, 37
Hyperbolic trigonometric functions, 49

I

identity() function, 79
Identity matrix, 79
if() function, 51, 52, 62
If statement, 161
Indefinite integrals, 194, 198, 204, 208
Index subscript, 15
Initializing an array, 68
Input, 148, 152–154
Input source, 152
Input table, 74
Insert bar, 14
Inserting a minus sign, 16
Inserting a row or column into an existing array, 80
Integrals, 4
Integration
    without curve fitting, 225
    via curve fitting, 225
intercept() function, 122, 132
interp() function, 119–121, 132
Interpolation, 118–122, 139
Inverse, 192
Inverse hyperbolic trigonometric functions, 49
Inverse operation, 192
Inverse trigonometric functions, 48
Inversion, 89
Irrigation system
    designing, 41
Iteration variable, 166
Iterative solutions, 217
Iterative solve block, 218
Iterative solver, 217

K

Keywords
    conventions used in text, 6–7
Kronecker delta, 52

L

Laplace transforms, 4
last(v) function, 94, 115
Law of refraction, 48
length(v) function, 94, 122, 129
lgsfit(vx, vy, vg), 129
Limitations to Mathcad's units capabilities, 23
Linear models, 126
    exponential, 128, 129
    intercept, 122
    logarithmic, 129
    logistic, 129
    power, 129
    sine, 129
    slope, 122
Linear regression, 122
    generalized, 124
linfit() function, 124, 129
linterp() function, 118–119, 132
Live symbolic operator, 148, 182
ln(z), 45
Load cells, 140
Local definition, 143
Local definition operator, 155
Local definition symbol, 149
Local scope, 155
Local variables, 144
log(z), 45
Logarithm and exponentiation functions, 45–47
Logarithms, 129
logfit(vx, vy, vg), 129
    Logical condition, 161
Logistic, 129
Loops, 148, 164–169
    break statement, 167–168
    continue statement, 168–169
    for loops, 166–167
    iteration variable, 166
    structures, 164
    while loops, 165–166
lspline(), 119–121, 132
Luminosity, 21

M

Manipulating Equations, 186
    collecting terms, 186
    expanding, 186, 187

factoring, 186, 187
partial-fraction expansion, 186, 190
Mass, 21
Math operators, 44, 45, 154
Math Toolbar, 10
    Symbolic Keyword Toolbar, 184, 186, 206
Mathcad programs, 142
Mathcad
    anchor point, 11
    assignment operator (:=) 12
    automatic recalculation, 3–4
    case sensitivity of, 53
    curve fitting, 122–131, 236
    defining new unit, 21
    design tool, use as a, 3–4
    displaying a value, 12
    editing session, 29, 30
    engineer's tool kit and, 5–6
    entering/editing text, 27–29
    equation
        changing an operator, 16–20
        changing a value/variable name, 16
        entering, 14–20
        selecting part of, 15
    equation edit mode, 13
    Font Bar, 10
    Format Bar, 10
    Format Menu, 10
    functions, 42–45, 47–56, 94–96
        defined, 46, 237
        mean() function, 46
        scalar functions, 43, 44
        user-written functions, 47–49
        uses of, 46–47
    fundamentals of, 9–32
    global assignment, 12
    graphing with, 110–115
    index subscripts, 15
    information sharing and, 79
    interpolation, 118–122, 139
    mathematical problem solver, use as a, 4
    Math Toolbar, 10
    Menu Bar, 10
    operator precedence rules, 14
    order of solving equations in, 10
    order for presenting results, 5
    problem-solving ability, 1–3
    programming, 141–180
    QuickPlot, 3–4, 114, 115
    regions
        View Menu, 10
    statistical functions, 117, 118
    styles, 10
    symbolic equality, 12
    symbolic math capabilities, 181, 182, 184, 186–192, 194–195, 198, 201
    text subscripts, 15
    Title Bar, 10
    unit converter, use as a, 4–5
    units, 20–25
        abbreviations, 21
        capability limitations, 23
        conversions, 22
        defining a new unit, 22
        editing on a value/result, 23
        placeholders, 21, 22
    View menu, 10
    worksheets, 3, 10
Mathematical operators, 44–45
Mathematical problem solver
    Mathcad as, 4
Matrices, 25, 66–100
    arguments, 67
    array origin, 68
    arrays, 72, 94
        choosing a subset of, 86
        combining, 86
        copying/pasting from other Windows programs, 78–79
        deleting rows/columns from, 82
        functions, 42–45, 47–56, 94–96
        initializing, 68–79
        inserting row/column into, 80, 82
        matrix() function, 77
        maximum size of, 68
        modifying, 79–87
        operations, 87–94
        range variables, 77
        reading values from a text file, 70
        selecting a portion of, 82
        selecting a single column of, 86–87
        selecting a single row of, 85
        two-dimensional array, 77
        typing values into, 68
        units on matrix elements, 77
    defined, 67–68, 237
    determinant, 91–93
    display, 25–26
    element-by-element multiplication, 88, 89
    elements, 67
    identity matrix, creating, 79
    inversion, 77, 89
    max(A) function, 94
    min(A) function, 94
    multiplication, 88
    operations, 4
    sorting, 94
    Toolbar, 10
    transposition, 89
    vector, 67
matrix( ) function, 77
max(A) function, 94
Maximum Array Size, 68
mean() function, 53, 117
mean(A), 117
Median
    finding, 106
Menu Bar, 10
Menu selections
    conventions used in text, 7
min(A) function, 94
Minus sign, 16
Mixed limits
    symbolic evaluation with, 198
$mod(x,y)$, 50
Modifying graphical display, 115
Modulo division, 50
Moving text, 29
Multiloop circuits, 104
Multiple loads, 63, 64

N

Nanotechnology, 8, 32
Newton's law, 29
Nondecimal values
    entering, 14–17, 19, 20
Nth Derivative button
    Calculus Toolbar, 203, 204, 209, 193
Nth root operator, 44
num2str(x), 53
Numeric limits
    symbolic evaluation with, 198
Numerical differentiation, 233–237
    derivatives from experimental data, 234, 236, 237
    evaluating derivatives of functions numerically, 234
Numerical integration, 222, 223, 225–227, 229, 232, 233
    data sets, 227, 228
        integration via curve fitting, 192
        integration without curve fitting, 227–229
    Simpson's rule, 229, 232
Numerical techniques, 216–224, 226, 227, 229, 232–234, 236–245
    iterative solutions, 217–222
    numerical differentiation, 234, 236, 237

numerical integration, 222, 223, 225–229, 232, 233

O

Objectives of the text, 6
Octal, 14
On error statement, 164
Operations, 148, 154
Operators, 44–45, 154–155
    changing the displayed symbol, 17–19
    precedence rules, 14, 155
Optics, 6, 42–43, 54–55
Orifice meter calibration, 136
ORIGIN, 68
Otherwise statement, 163–164
Output, 148, 156–161

P

Parameters, 43, 143, 150
Partial fraction expansion, 190–191
Passing arguments
    by address, 151
    by value, 151
Percent (%), predefined variable, 13–14
Physical property data fitting, 128, 138
Pi (π), predefined variable, 13–14
Placeholders, 21, 22
Plotting, 110
    multiple curves, 112
    vector against vector, 110–113
Polynomial coefficients, 190–191, 193, 207
polyroots() function, 219
Populations, 118
Power, 21, 129
Practice, 9
Predefined variables, 13–14
Pressure, 21
Product operator, 44
Programming in Mathcad, 141–180
    Conditional execution, 148
    Data, 148
    Functions, 148,
        See also Function, Functions
    Input, 148
    Loops, 148
    Operations, 148
    Output, 148
Programming Toolbar, 10, 144–145, 148
Program name, 143
Program region, 142
pspline() function, 119–121, 132
pwrfit(vx, vy, vg), 129

Parameters, 43

Q

QuickPlot, 3, 4, 46, 114, 115
    and visualization of functions, 47

R

R (unit), 21
R^2, coefficient of determination, 122
Radians, 47
Radix, 14
Random-number-generating function, 43, 52
Range variables, 76, 77
READ () function, 53
Reading values from a file, 70
READPRN () function, 53, 70, 152
Regions
    View Menu, 10
Regression, 4
Residual plots, 123
Resistance temperature devices, 105
Results
    display of, 24–27
    documenting, 29
    inserting into other software programs, 5
Returning multiple values, 158
Return statement, 158
Return value, 142
reverse(v) function, 94
Right arrow symbol, 182
Risk analysis, 6, 109, 110
Risk management, 109
root( ) function, 219
Round-off and truncation functions, 50, 51
rows(A) function, 94
rsort() function, 94
rsort(A,n), 95

S

Samples, 118
Scalar functions, 43, 44
    advanced math functions, 50, 51
    elementary math functions and operators, 44, 45
    file-handling functions, 53
    string functions, 53
    trigonometric functions, 47–50
sech(z), 49
sec(z), 47
Selecting part of an equation, 14

Selecting text, 29
SI, 20
Simple linear regression, 122–124, 135
Simplify, 189
Simplify operation, 187–190, 207
Simpson's rule, 229, 232
Simpson's-Rule Integration, 229
Simultaneous equations, 98, 99
sin(z), 47
Sine, 129
sinfit(vx, vy, vg), 129
sinh(z), 49
Sizing text, 29
slope, 122
slope() function, 122, 132
Solids storage tank
    relating height to mass for, 140
Solve block, 218
Solving an equation symbolically, 185
sort(v) function, 94
Sorting, 94
Spline interpolation, 118
Spreadsheets, 5
Spring constants, 36, 37
Square root operator, 44
stack( ) function, 86
Standard trigonometric functions, 48
Statistical calculations, 4
Statistical functions, 117, 118
Statistics, 138
stdev(A), 117
Stdev(A), 118
Steady-state conduction, 107
Storage tank substance
    measuring volume/mass of, 61–63
str2num(s), 53
Stress-strain diagrams
    analyzing, 205, 206
    replotting as a force-displacement graph, 206
String functions, 53
strlen(s), 53
Styles, 10, 29
Styling text, 29
submatrix() function, 86
Subscripts, 28
Substance, 21
Substitute, 188
Substitution operation, 187–189
substr(S,n,m), 53
Summation operator, 44, 227
Superscripts, 28
Suspension bridge
    designing, 41, 64
Symbolic differentiation, 201
Symbolic equality (=), 12

Symbolic integration, 193, 194, 195, 198, 199, 201, 202
    definite integrals, 198, 199, 201–203, 208
    differentiation, 203–205
    indefinite integrals, 194, 208
    multiple variables with, 195–197
    indefinite integral operator and, 198, 201
    Symbolics menu and, 198–201
    work, 206
    Young's modulus, 205, 206
Symbolic Keyword toolbar, 10, 184–186, 206
    operations using, 207–209
    substitution using, 188, 189
Symbolic math, 182
Symbolic math functions, 181–183, 185–195, 197–199, 201
    definite integrals, 198, 199, 201–203, 208,
    locations of, 184–186
    manipulating equations, 188–190, 192
        collect operation, 190, 207
        expand operation, 187, 188, 190, 207
        factor operation, 187, 188, 207
        partial fraction expansion, 187–193
        simplify operation, 187–190, 207
        substitute operation, 154–189
    polynomial coefficients, 191, 193, 207
    solving in equations symbolically, 186, 187
    symbolic integration and differentiation, 194, 195, 198, 201–205
        definite integrals, 198, 199, 201–203, 208
        differentiation, 203–205
        indefinite integral operator and, 198–201
        indefinite integrals, 194, 208
        multiple variables and, 195, 197
        Symbolics menu and, 198–201
        work, 206
        Young's modulus, 205, 206
    Symbolic Keyword Toolbar, 184–186, 206
    Symbolic matrix math, 191–193
    Symbolics menu, 185, 186
Symbolic matrix math, 192–193
    determinant operation, 193

    inverse operation, 192
    transpose operation, 192
Symbolics menu
    substitution using, 188, 189
    symbolic integration using, 198–201
Symbolics menu, 182
Systems of units, 20–25

T

$\tan(z)$, 47
$\tanh(z)$, 49
Temperature, 21
Temporary variables, 155
Tension and angles, 64
Text
    entering/editing, 27–29
    selecting, 28
    sizing/moving, 29
    styling, 29
    sub-/superscript, 28
Text edit mode, 27
Text region, 27–29
Text subscript, 15
Thermocouple calibration, 136
Time, 21
Title Bar, 10
Tolerance, 218
Toolbars, 10
Total recycle, 6, 66, 95
trail() function, 51, 53
Transit, 51, 237
Transpose, 192
Transpose operation, 192
Transposition, 89
Trapezoidal-rule integration, 227
Trial-and-error calculations, 217
Trigonometric calculations, 4
Trigonometric functions, 47–50
    hyperbolic, 49
    inverse, 48
    inverse hyperbolic, 49
    round-off and truncation functions, 50, 51
    standard, 47
Two-dimensional array
    example of, 77
Typing values into an array, 68

U

Unit conversions, 19
Unit converter
    Mathcad as, 4–5
Unit names, 21

Units, 20–26
    abbreviations, 21
    cautions on names, 21
    CGS system, 20
    conversions, 22
    defining new, 22
    display, 26–27
    editing on a value or result, 24
    Mathcad capability limitations and, 23
    MKS system, 20
    matrix elements and, 76–77
    names, 21
    no system, 20
    placeholders, 21, 22
    SI system, 20
    simplifying, 186
    substituting, 186
    systems of, 20–25
        dimensions supported by, 21
    US system, 20–25
User-written functions, 47–49, 53
Using an input table, 74

V

Value
    changing, 15–17, 19, 20
Vapor pressure, 60, 61
Vapor-liquid equilibrium (VLE), 137
var(A), 117
Var(A), 118
Variable names
    changing, 15–17, 19, 20
    conventions used in text, 6–7
Vector, 67
    plotting vector against, 110–113
Vector Toolbar, 10
Vectorize operator, 88
View menu, 10
Volume, 21

W

Wheatstone bridge, 104–106
While loops, 165–166
Work, 21, 206
Worksheet, 3, 10, 11
    validating while developing, 47
WRITE() function, 53
WRITEPRN() function, 53

Y

Young's modulus, 205, 206